IMAGES
of America

NORTHEAST FOREST FIRE SUPERVISORS

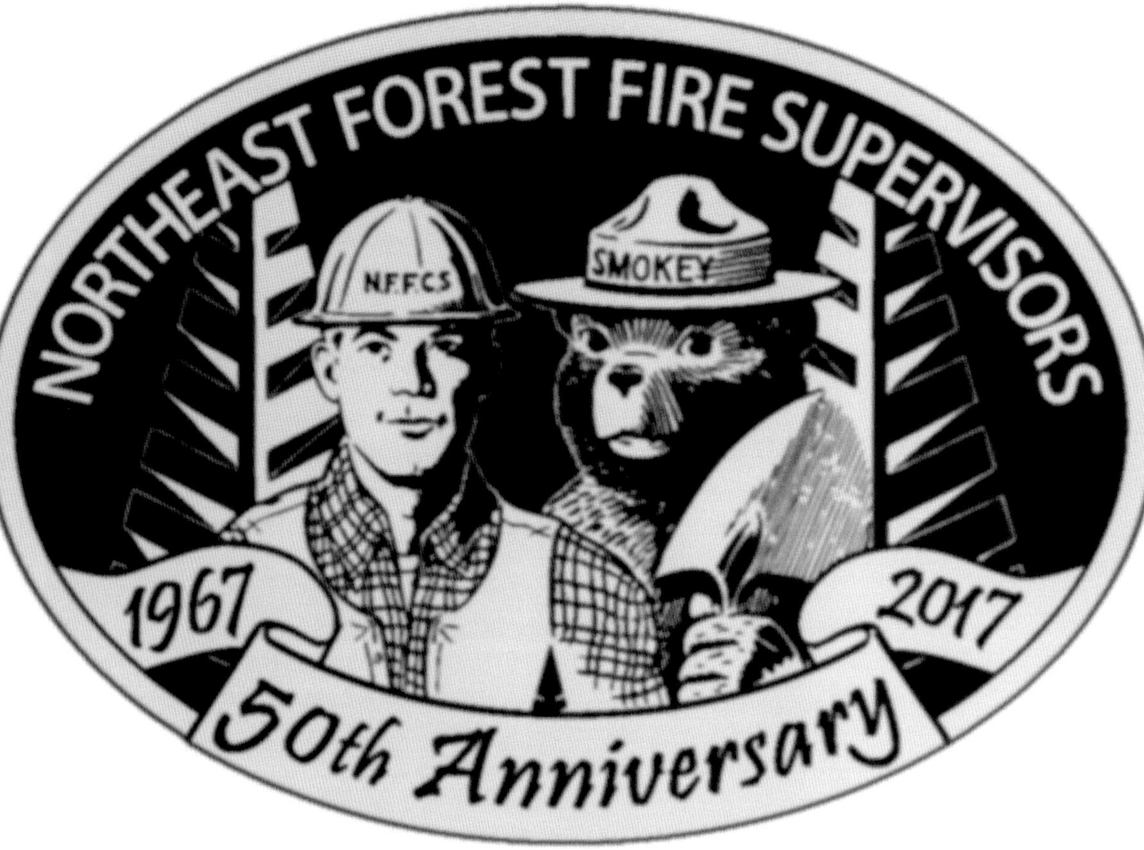

The Northeast Forest Fire Supervisors (NFFS) was established in 1966, and their first official meeting was held in Philadelphia, Pennsylvania, in 1967. To commemorate the 50th meeting, members updated the logo to highlight the initials of the organization's first name NFFCS, Northeast Forest Fire Control Supervisors, inscribed on the helmet. A fire supervisor is the lead individual in a state forestry program (or equivalent) who oversees wildland fire control and fire management activities. (NFFS.)

On the Cover: This 1962 photograph captures a Jeep firefighting unit and several Maryland Forest Service firefighters; they are demonstrating the use of equipment and water in containing wildfires. During that era, most state forest firefighting agencies across the Northeast took advantage of readily available vehicles and equipment and developed them into necessary mobile firefighting units. Around the time of the photograph, the Maryland Forest Service firefighting fleet consisted of 1.5-ton trucks, bulldozers with fire plows, and all-terrain Jeeps. (Maryland Forest Service.)

IMAGES
of America

NORTHEAST FOREST FIRE SUPERVISORS

Northeastern Forest Fire
Protection Compact

ARCADIA
PUBLISHING

ISBN 978-1-4671-2461-4

Published by Arcadia Publishing
Charleston, South Carolina

Printed in the United States of America

Library of Congress Control Number: 2016959077

For all general information, please contact Arcadia Publishing:
Telephone 843-853-2070
Fax 843-853-0044
E-mail sales@arcadiapublishing.com
For customer service and orders:
Toll-Free 1-888-313-2665

Visit us on the Internet at www.arcadiapublishing.com

CONTENTS

ACKNOWLEDGMENTS

The Northeastern Forest Fire Protection Compact would like to thank all 20 northeastern state forest fire supervisors, as well as their staffs, who participated in providing many exceptional images and historical documentation toward the development of this commemorative book. A big thank-you goes to the US Forest Service's Northeastern Area State & Private Forestry for supporting this project and the 50th anniversary of the Northeast Forest Fire Supervisors. A wealth of gratitude is expressed to Maris Gabliks for providing technical assistance in the coordination of the book's development, researching the history of the Northeast Forest Fire Supervisors, and working with all 20 states. The following people all helped contribute stories, photographs, and historical information for inclusion in the book. They include Tom Parent, John Berst, Catherine Koele, Steve Creech, Trent Marty, Rich Schenk, John Peterson, Gail Kantak, Tom Wilson, Bev Stout, Lindsay Brown, Christi Powers, Trevor Augustino, Dave Celino, David Robbins, Bill Hamilton, Mimi Barzen, Dan Munn, Jim Fisher, Ben Webster, Steve Sherman, Andy Jacob, Aaron Kloss, Levi Gelnett, Olney Knight, Lars Lund, Tom Brady, Matt Dillon, Ron Stoffel, Brad Simpkins, Scott Bassett, Tom Muise, and Walt Jackson.

The images in this book appear courtesy of the following state and federal forest fire protection agencies and organizations: Connecticut Division of Forestry; Delaware Forest Service; Iowa Department of Natural Resources (DNR) Bureau of Forestry; Illinois DNR Division of Forestry Resources; Indiana DNR Division of Forestry; Massachusetts Bureau of Forest Fire Control; Maryland Forest Service; Maine Forest Service; Michigan DNR Forest Resources Division; Minnesota DNR Division of Forestry; Missouri Department of Conservation (DOC) Forestry Division; New Hampshire Division of Forests and Lands; New Jersey Forest Fire Service; New York Department of Environmental Conservation (DEC) State Forest Rangers; Ohio DNR Division of Forestry; Pennsylvania Department of Conservation and Natural Resources Division of Fire Protection; Rhode Island Division of Forest Environment; Vermont Department of Forests, Parks, and Recreation; West Virginia Division of Forestry; Wisconsin DNR Division of Forestry; Northeastern Forest Fire Protection Compact (NFFPC); Mid-Atlantic Interstate Forest Fire Protection Compact (MAIFFPC); Great Lakes Forest Fire Compact (GLFFC); Big Rivers Forest Fire Management Compact (BRFFMC); US Forest Service (USFS); Michigan Roscommon Equipment Center (REC); and Northeast Forest Fire Supervisors (NFFS).

Introduction

Managing wildfires has been a significant mission across the nation since before the turn of the 20th century. Each state has had its own specific fire management issues, but there always has been a bond among the states by working collaboratively and sharing thoughts and ideas that benefited them all. This book will concentrate on telling the story through photographs of the 20 unique forest fire control and management agencies that protect the northeastern area of the United States.

Many of these agencies have already reached their centennial celebrations and have strong histories in protecting life, property, and natural resources from wildfire. These northeastern states have worked together independently, but their cooperative endeavors within their forest fire compacts and with the US Forest Service, through wildfire suppression and management mutual aid agreements and federal Cooperative Fire Protection funding, have developed a key network and partnership in delivering excellent fire protection programs to the citizens of their states and commonwealths.

The first federal cooperation with the states in protecting public and private forestlands from fire occurred on March 1, 1911, with the creation of the Weeks Act. This law allowed states to cooperate with any other state or states, as well as with the US government, regarding fire protection. The law also provided federal funds to states to provide fire protection in forested watersheds of navigable streams if the state had created a forest fire protection law and the funds provided by the federal government did not exceed the amount of funds provided by the state. In 1924, the Clarke-McNary Act authorized the Department of Agriculture to cooperate with states in developing systems for protecting nonfederal lands from fire and expanded the protected area to all timbered or forest-producing lands that were protected by the state or private landowners. Since the enactment of these federal laws, the cooperative endeavors between the US Forest Service and 20 northeastern states has continued to grow and prosper. In 2000, the National Fire Plan was enacted and provided federal funds to the states to accomplish priority fire management projects across the area. Today, the National Cohesive Strategy works to unite local, state, and federal partners to restore and maintain resilient landscapes, create fire-adapted communities, and improve wildfire response.

After a series of disastrous wildfires across Maine in 1947, the first subregional interstate cooperation involved the creation of the Northeastern Forest Fire Protection Compact in 1949. Another compact formed was the Mid-Atlantic Interstate Forest Fire Protection Compact, established in 1965. The Great Lakes and Big Rivers Compacts were founded in the 1980s. These compacts have allowed the state forest fire supervisors to interact on a subregional basis across the 20 northeastern states. These compacts continue to play an integral part in coordinating wildfire response across the Northeast, as well as providing leadership in critical training and fire management–related activities among all 20 states and partner agencies. The compacts in the Northeast have also recently played a key role in the development of an alliance of forest fire compacts that works to share information and coordinate the activities among all compacts in Canada and the United States.

Prior to the Northeast Forest Fire Supervisors organization being created in 1966, early coordination between the states and US Forest Service's Cooperative Fire Control program occurred during forest firefighting equipment demonstrations. These demonstrations were arranged by groups of states and the US Forest Service in the north central and eastern regions to share information, technology, and developments in fire suppression. A 1947 US Forest Service report documented one such meeting that was sponsored by the Northeastern States Forest Fire Equipment Committee in 1947 and occurred in New Jersey with equipment showcased by the states of Maryland, New York, Pennsylvania, Connecticut, and New Jersey. An article in the US Forest Service's *Fire Control Notes* captured another equipment demonstration hosted by the Great Lake states in 1946 that provided field tours in Michigan and Wisconsin with equipment demonstrations from both states, as well as Minnesota.

In 1965, the secretary of agriculture, Orville L. Freeman, established a review team to study the US Forest Service's organization. This report, referred to as the Deckerd Report, resulted in reorganizing US Forest Service regions in the East and basically created the Northeastern Area State & Private Forestry unit. This new Northeastern Area State & Private Forestry unit concept joined all 20 northeastern states in their cooperative work with the US Forest Service. As a result of this new US Forest Service structure, the forest fire supervisors working in those 20 states realized that an organization of their peers was necessary to be aligned with the US Forest Service. The Northeast Forest Fire Supervisors (NFFS) was organized in 1966, and its first meeting was held in Philadelphia, Pennsylvania, on January 17–19, 1967. The group was originally named the Northeast Forest Fire Control Supervisors and was composed of the forest fire control supervisors from the 20 northeastern states. These 20 forest fire supervisors have interacted on a routine basis within their sub-geographical areas based on the needs of delivering their fire protection programs and dealing with issues of mutual concern. The Northeast Forest Fire Supervisors meet annually in a different location that highlights the fire program of the hosting state entity. Annual meetings cover relevant topics regarding wildfire management and cooperation among the 20 states and with the Forest Service and other federal partners. The states involved are as follows: Maine, New Hampshire, Vermont, Rhode Island, Connecticut, Massachusetts, New York, Delaware, Maryland, New Jersey, Pennsylvania, West Virginia, Ohio, Virginia, Michigan, Minnesota, Wisconsin, Illinois, Iowa, Missouri, and Indiana. (Virginia is not within the northeastern states but is a member of the Mid-Atlantic Interstate Forest Fire Protection Compact and has been included here.) The NFFS has strong traditions since its establishment, and one of the annual highlights is to honor an individual from among the 20 states who makes significant contributions toward forest fire protection by awarding him or her with the annual Northeast Forest Fire Supervisors Award.

The images in this book have captured a sample of the moments in the 100-plus year history of the 20 northeastern state wildfire protection agencies and the history of the four forest fire compacts. These moments have been combined and highlighted by the overarching partnership developed through the Northeast Forest Fire Supervisors organization over the last 50 years. The Northeast Forest Fire Supervisors has been instrumental in bringing together 20 states, four forest fire compacts, and their federal partners to achieve a coordinated and highly professional wildland fire management program.

One

Northeast Forest Fire Supervisors

The Northeast Forest Fire Supervisors (NFFS) was organized in 1966, and their first meeting was held in Philadelphia, Pennsylvania, on January 17–19, 1967. The organization was originally named the Northeast Forest Fire Control Supervisors and represented the forest fire control supervisors from the 20 northeastern states. The original purpose of the organization was to stimulate and improve greater overall efficiency in wildland fire protection of rural areas and wildlands. The organization was composed of state and federal forest service members who were charged with the responsibility for stimulating and promoting the development and use of specialized forest fire equipment, better techniques in fire prevention, pre-suppression and suppression, improved training, and safety methods. The overall goal was to attain and maintain the least number of fires, with the least damage, and at minimal suppression cost. A fire supervisor is the lead individual in a state forestry program (or equivalent) that oversees wildland fire control and fire management activities. The original executive committee consisted of three state forest fire supervisors, the US Forest Service's chief of fire control for the eastern region, and the chief of fire control from the Northeastern Area State & Private Forestry office.

In the mid-1970s, the members of the organization changed its name to the Northeast Forest Fire Supervisors (NFFS) to better reflect their responsibilities in forest fire management beyond just the scope of controlling wildfires. In 2006, the NFFS became a committee of the Northeast Area Association of State Foresters to better coordinate activities among both organizations and other forest-related committees operating in the northeastern states. Over the years, the Northeast Forest Fire Supervisors has played a role in developing some significant wildland fire management programs in its region, such as the Roscommon Equipment Center in 1971, the Northeast Area Training Office concept in 1984, and the Northeast Area States Aviation Committee. All these initiatives have had national implications in wildfire management. The NFFS have had various working committees since 1967, which have included public information, film, research, equipment development and testing, railroad, training, fire prevention, wildland urban interface, mobilization, and aviation topics. During the past 50 years, the NFFS have been and continue to be leaders in addressing wildland fire protection issues that impact the northeastern parts of the United States.

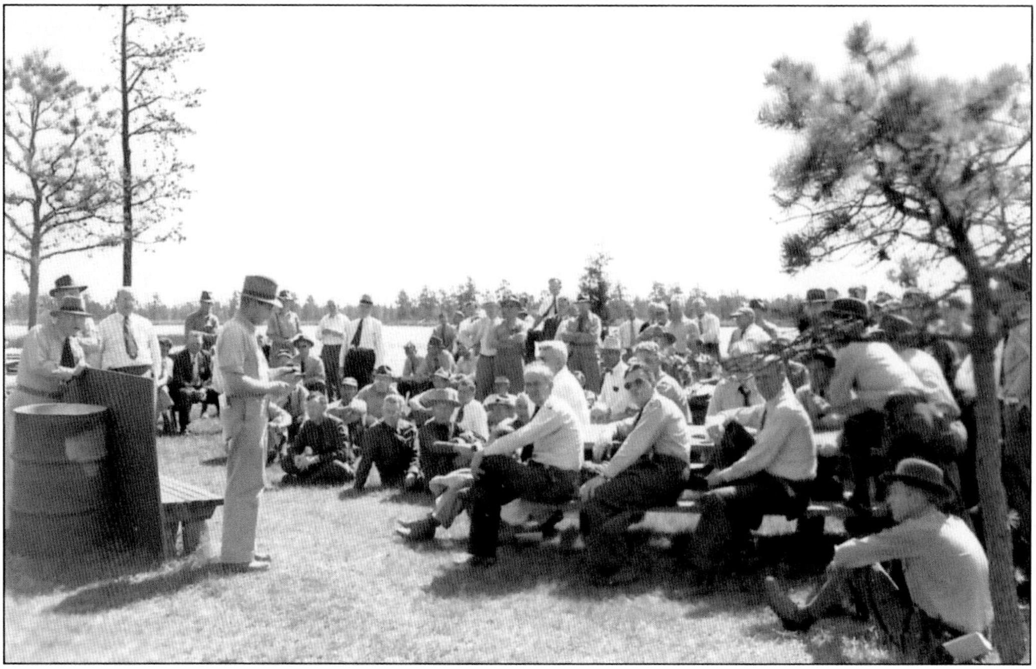

The US Forest Service (Region No. 7) and the Northeastern States Forest Fire Equipment Committee hosted their first forest fire equipment demonstration in New Jersey on September 18, 1947. Ernie Karger, chairman of the Northeastern States Forest Fire Equipment Committee, delivered a presentation during lunch at Penn State Forest. One hundred twenty-eight individuals were in attendance from 13 states and 2 US National Forests, Jefferson and George Washington. Many states brought pieces of equipment to demonstrate, which fostered an exchange of ideas toward developing better equipment. (USFS.)

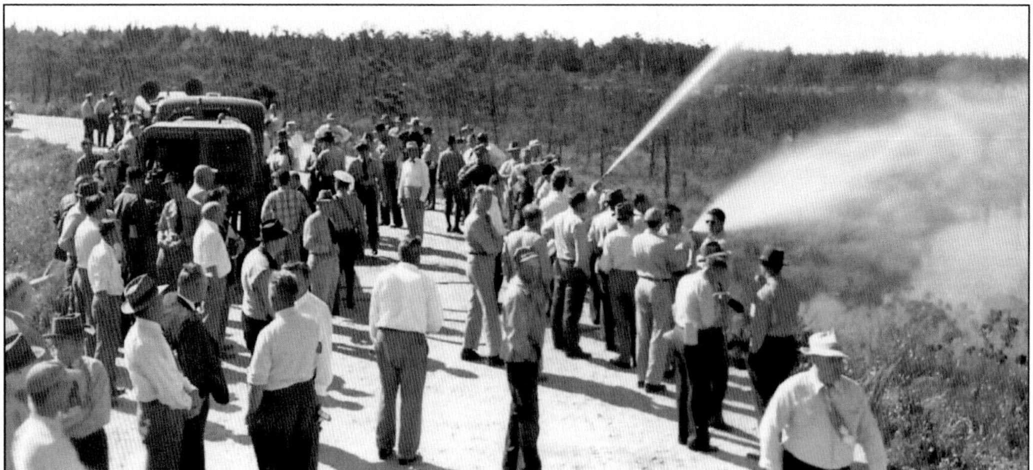

The Connecticut Forest Service demonstrates the use of homemade fog and spray nozzles at the first Northeastern States Forest Fire Equipment Committee demonstration in 1947. These nozzles were constructed for less than $1 by using a brass pipe cap with a quarter-inch orifice and perforated baffle plate. Attendees reported that the streams appeared quite satisfactory. These types of sessions served as the predecessor for the development of the future Northeast Forest Fire Supervisors organization. (USFS.)

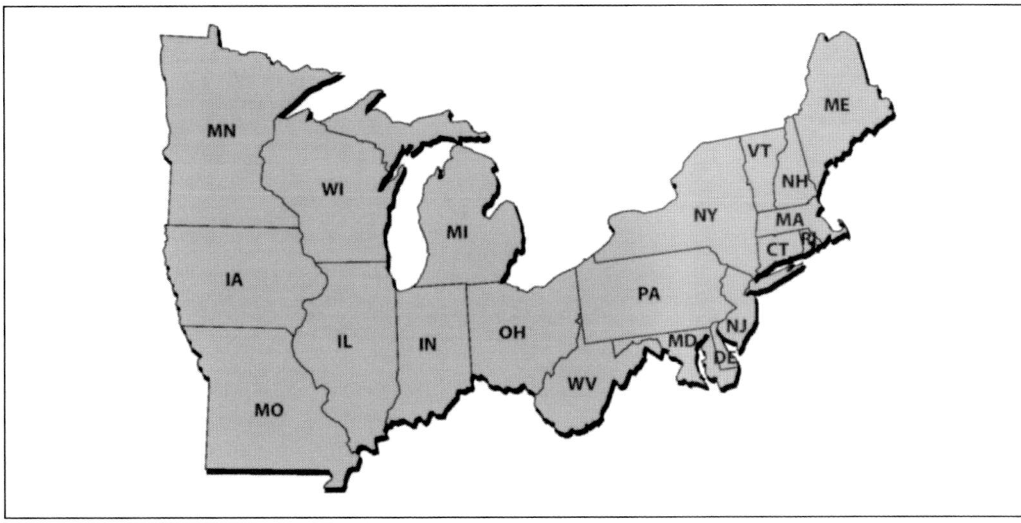

This map shows the states represented by the Northeast Forest Fire Supervisors. The Northeast Regional Action Plan for the National Cohesive Strategy reports that more than 40 percent of the land shown is forested, while 41 percent of the nation's population resides within this area. Seventy-four percent of the forestland is privately owned, while 26 percent is publicly owned. Available fire reporting data shows that, between 2002 and 2012, a total of 150,000 wildfires burned over 600,000 acres of open space across the 20 northeastern states. (USFS.)

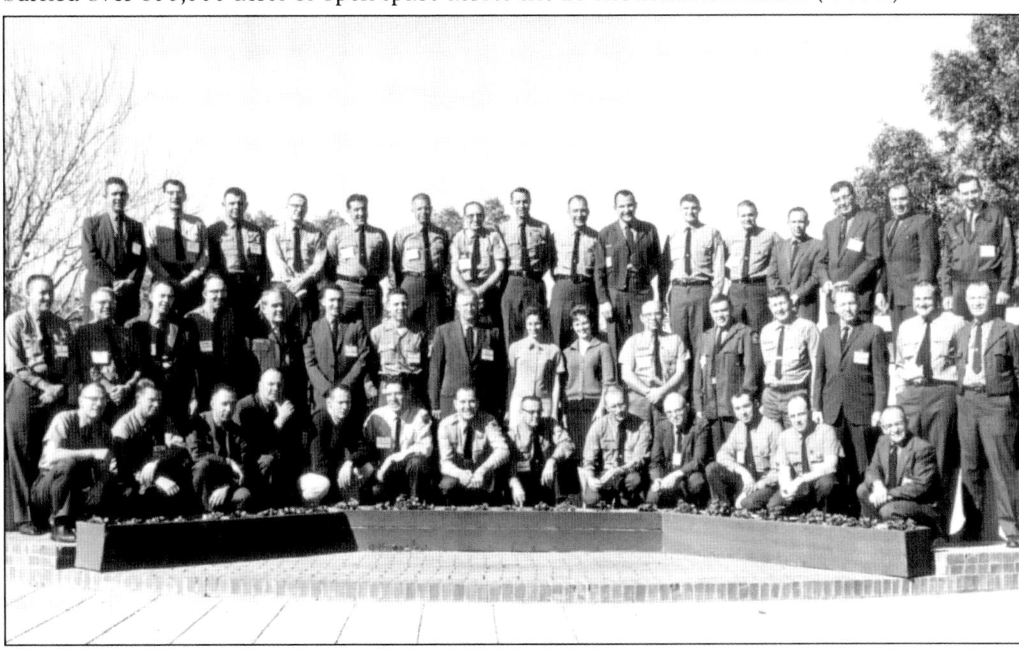

At an early meeting of the Northeast Forest Fire Control Supervisors, attendees included representatives from the 20 northeastern states, as well as representatives from the US Forest Service's Northeastern Area State & Private Forestry, US National Forests, and research stations serving the area. William Phoenix, of New Jersey, and Asher Kelly, of West Virginia, are among the forest fire control supervisors in attendance. Every meeting since 1967 has had a theme that described the main discussions and concerns of the meeting's agenda. (West Virginia Division of Forestry.)

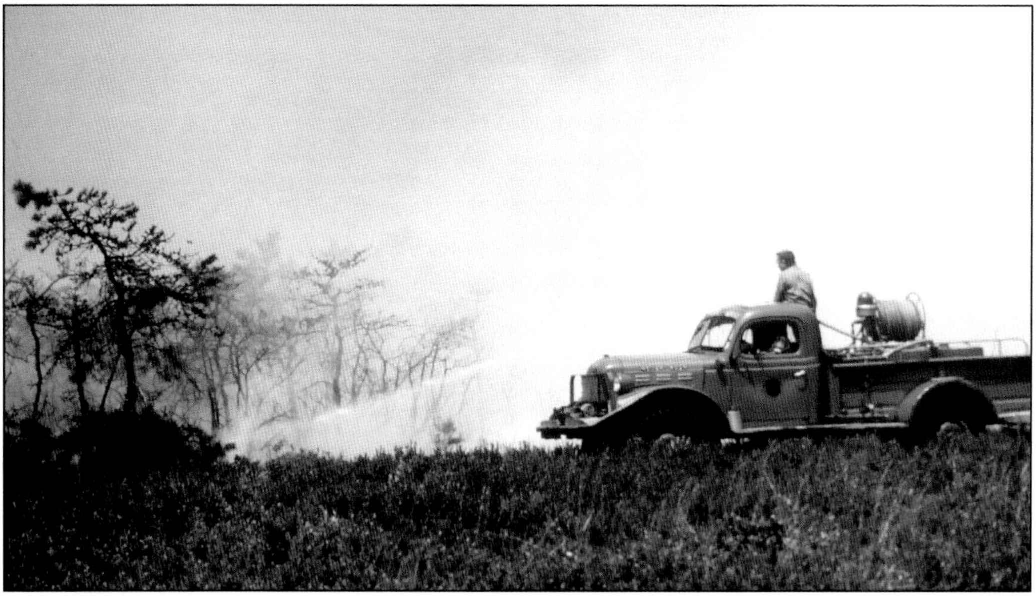

The NFFS have had an active film and training committee since its establishment. One of the films that the film and training committee helped develop was *Water vs Fire* in the late 1970s. The training film stressed wise water use and showcased different delivery methods, including engines, portable pumps, aircraft, and backpack pumps. The film was used to supplement the *Water vs Fire* handbook that was distributed by the Forest Service's Northeastern Area State & Private Forestry. (New Jersey Forest Fire Service.)

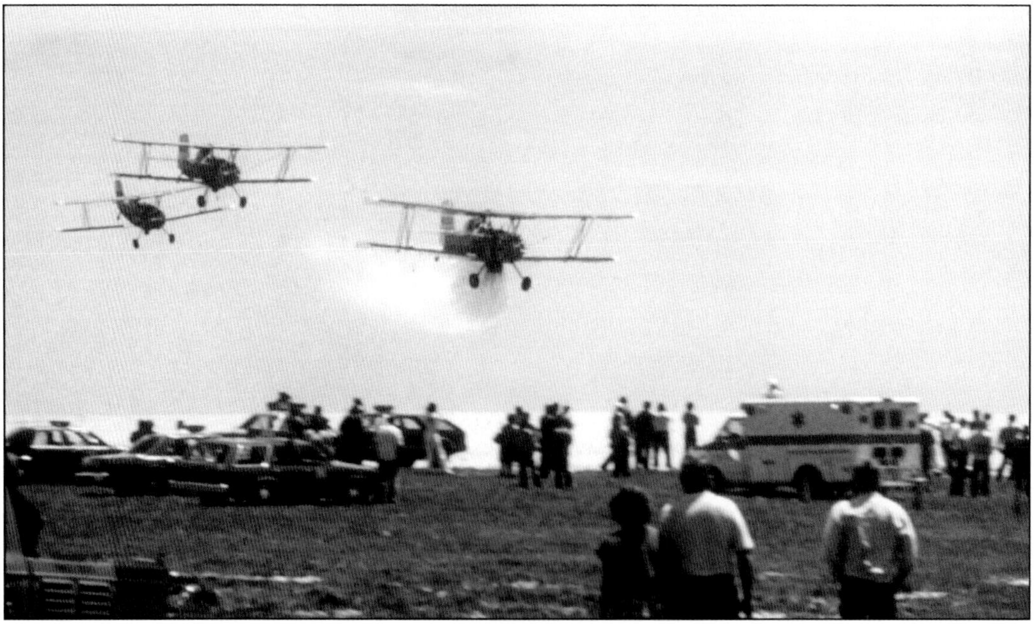

The 20th meeting of the NFFS was hosted by the New Jersey Forest Fire Service and included a field demonstration day at Coyle Field. The theme for the session was equipment and technology for the 1980s. As participants arrived at Coyle Field, they were greeted by three Ag Cat single-engine air tankers dropping red, white, and blue water in sequence on a burning target. (New Jersey Forest Fire Service.)

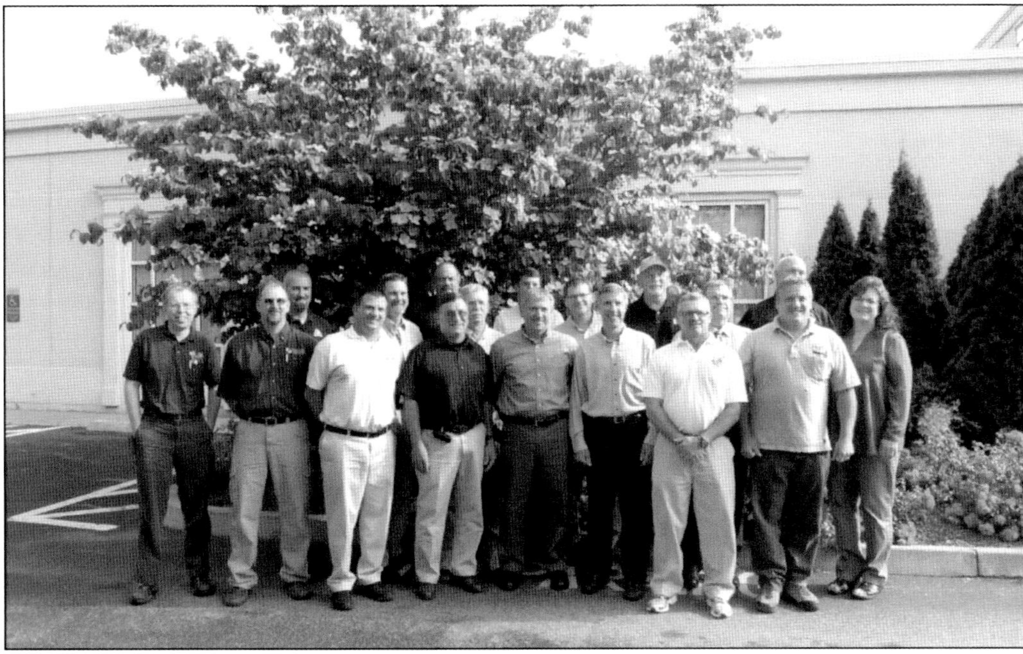

The 2012 Northeast Forest Fire Supervisors meeting was hosted by the state of Connecticut, and the meeting's theme was "Improving Fire Management Programs." The state forest fire supervisor representatives included the following: David Hall, Rhode Island; Monte Mitchell, Maryland; Ben Webster, Missouri; Mike Bowden, Ohio; Drew Daily, Indiana; Ralph Scarpino, Connecticut; Dave Celino, Massachusetts; Walt Jackson, West Virginia; Brad Simpkins, New Hampshire; Bill Hamilton, Maine; Lars Lund, Vermont; Scott Heather, Michigan; Ron Stoffel, Minnesota; Randy, Pennsylvania; Trent Marty, Wisconsin; Mike Drake, New Jersey; Tom Wilson; Illinois; and Gail Kantak, Iowa. (NFFS.)

Since the beginning, the Northeast Forest Fire Supervisors has been charged with the responsibility for stimulating and promoting the development and use of specialized forest fire equipment. Every other year's meeting consists of a field demonstration meeting where hands-on firefighting technology is shared among fire supervisors, their staff, and the manufacturers and suppliers of equipment. Drew Daily (Indiana, on left) and Tom Wilson (Illinois, on right) shoot water while firefighting pumps are demonstrating their output capabilities to the attendees. (NFFS.)

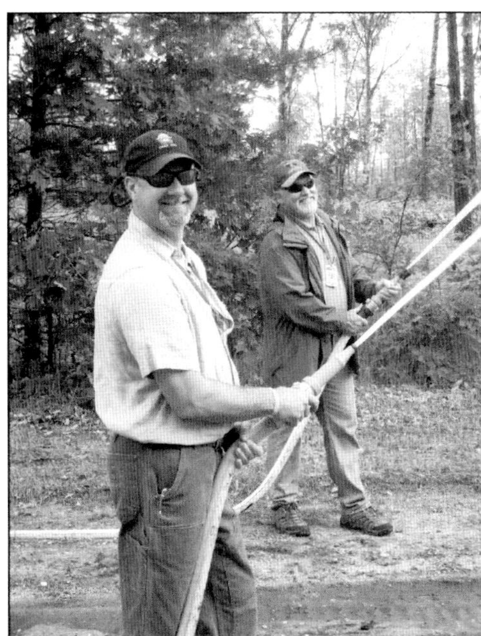

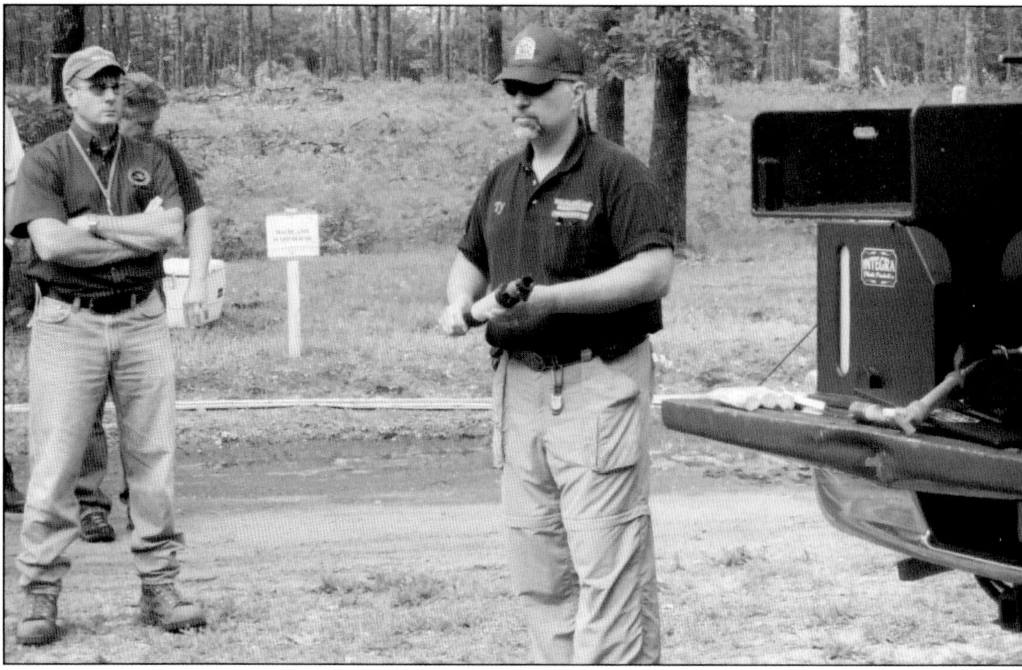

During the 2011 meeting in Michigan, Ty Parker, of Wildland Warehouse, provides an update on nozzles utilized in delivering foam to aid in the control of wildfires. Firefighting foam is one example of equipment, supplies, and gear utilized by wildland firefighters for many years, but the technology of the product, as well as new improvements in their delivery systems, makes it critical to get updates from researchers, manufacturers, and equipment distributors at the Northeast Forest Fire Supervisors meetings. The biennial equipment meetings have allowed this technology transfer to occur. (NFFS.)

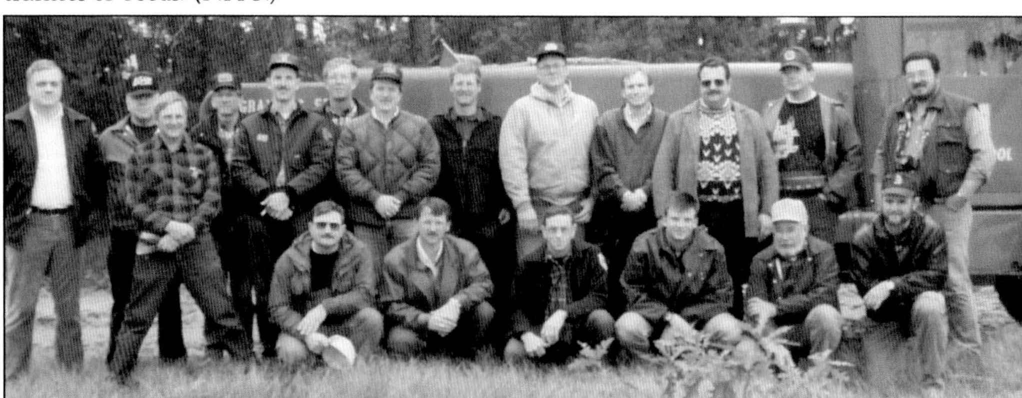

The Roscommon Equipment Center (REC) has held an annual mini workshop since 1974. The session allows participants to test equipment, interact one on one, and to tour the center where firefighting equipment is designed and produced. The 1993 participants included Kirk Bradley, Michigan Roscommon Equipment Center; Steve Frieswick, Massachusetts; Phil Luell, Wisconsin; Steve Maurer, New Jersey; Dick Mangan, US Forest Service; Gene Odata, Pennsylvania; John Dodge, New Hampshire; Bob Boyles, Ohio; Brian Hutchins, Michigan; Bill Hale, Missouri; Russ Arnold, Rhode Island; Hal Gaedtke, Wisconsin; Jim Carpenter, New York; Bill Hamilton, Maine; Bernie Barton, Vermont; Mike Bowden, Indiana; Mike Bailey, Wisconsin; and John Hadley, Ontario. (Ohio DNR Division of Forestry.)

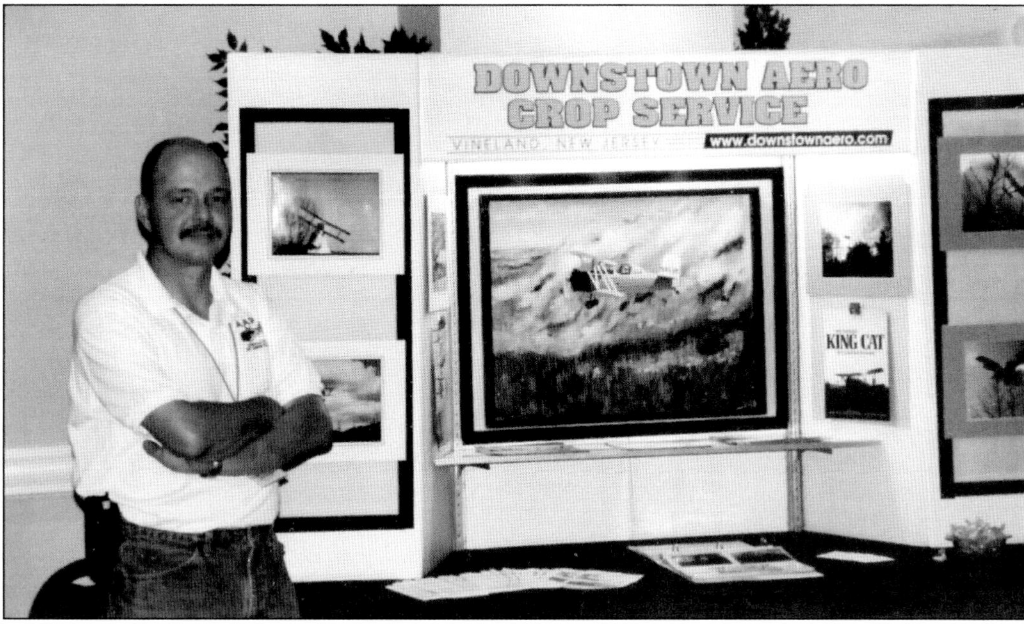

Equipment manufacturers and distributors have been a critical part of the annual Northeast Forest Fire Supervisors meetings. Improvements in equipment and new technology are showcased at the biennial hands-on demonstration meetings. This current information is utilized by fire supervisors in planning for their agencies' annual safety gear and equipment needs. In this photograph, Curt Nixhom stands beside a Downstown Aero Crop Service display at an annual meeting. (NFFS.)

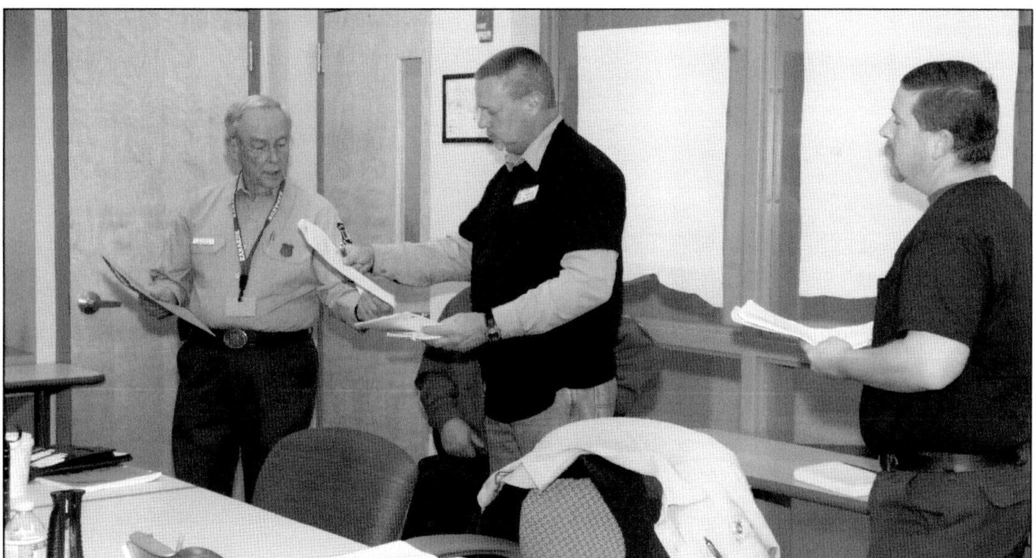

The concept of a Northeast Area Training (NEAT) office was instituted in 1984. Since then, NEAT has assisted and coordinated the delivery of hundreds of training courses across the 20-state area. NEAT continues to work closely with the forest fire compacts and partner federal agencies in providing a leadership role in training. In this photograph, US Forest Service's Northeastern Area State & Private Forestry fire director Billy Terry provides a briefing to a team completing the simulation portion of a S420 Command and General Staff course. (Ken Badger.)

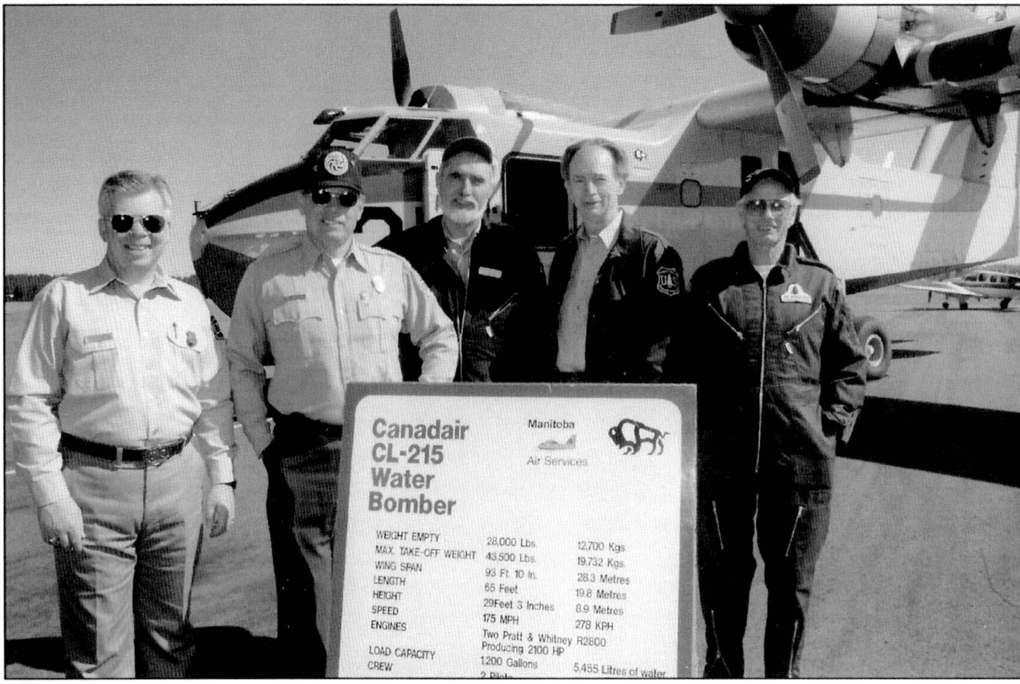

The Northeast Area States Aviation Committee (NASAC) was established in 1998 to provide aviation leadership across the 20 states. Most states have utilized aircraft during the history of their fire management programs whether state owned, Federal Excess Personal Property on loan, leased, or cooperatively used through agreements with partner agencies. This photograph captures, from left to right, Dan Zimmerman (US Forest Service); Ike Anderson, Minnesota DNR Division of Forestry; George Brooks (US Forest Service); and two contract pilots discussing the CL-215 water bomber program utilized by Minnesota DNR Division of Forestry. (Minnesota DNR Division of Forestry.)

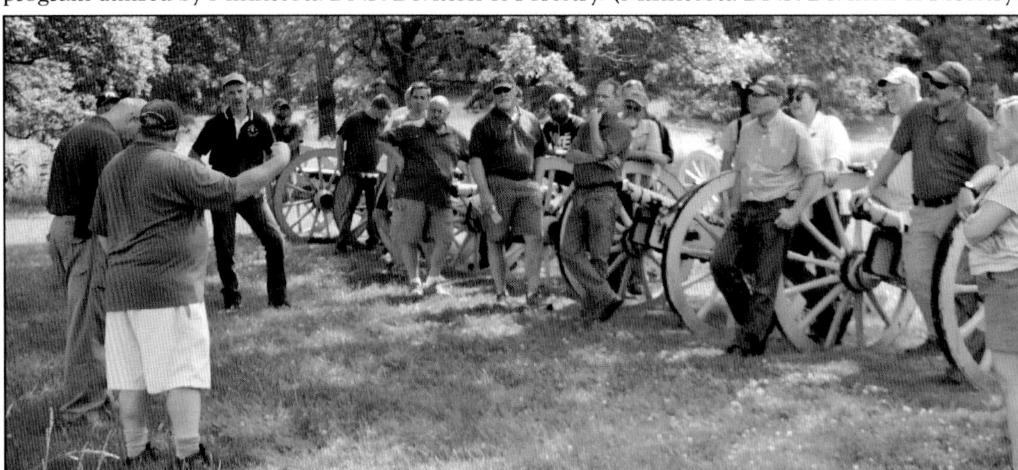

In conjunction with the 50th meeting of the NFFS, the US Army Combat Studies Institute's Staff Ride Team hosted a leadership staff ride at Valley Forge National Park. The Military Staff Ride was transformed for wildland fire leaders by adding wildland fire management and leadership issues that integrate well with military leadership and management. Topics included training, logistics, leadership, and facilities. The session was based on the Continental Army's encampment during the winter of 1777 in Valley Forge, Pennsylvania. (NFFS.)

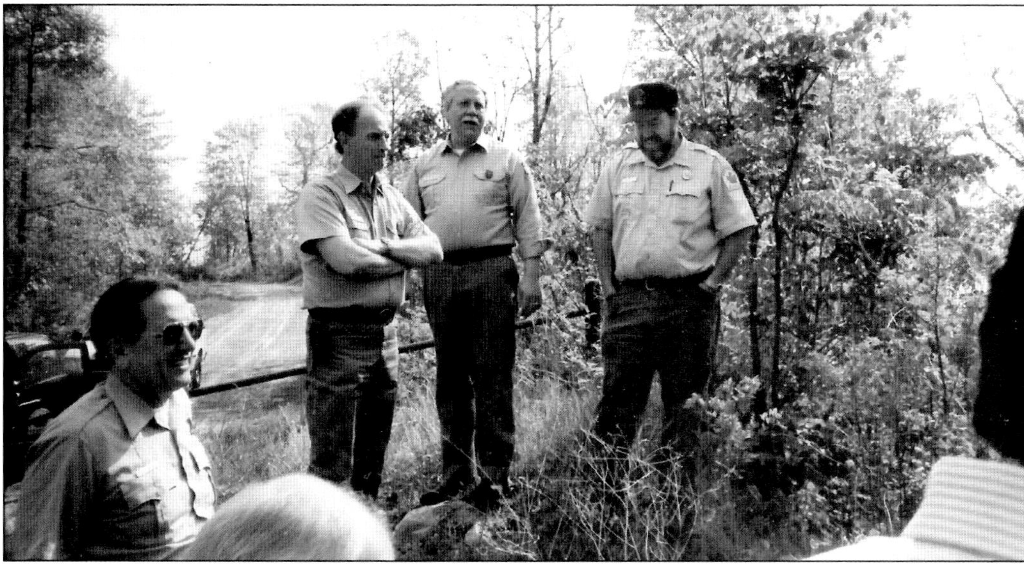

The Northeast Forest Fire Supervisors has had a strong relationship with the US Forest Service's Northeastern Area State & Private Forestry, as well as Region No. 9, since the organization's creation. This photograph captures US Forest Service and West Virginia Forestry personnel discussing wildfires caused by coal seam fires during a cooperative fire program review in 2001. From left to right are Steve Upton, George Brooks, Mac Gramley, and Kevin Arnold. (NFFS.)

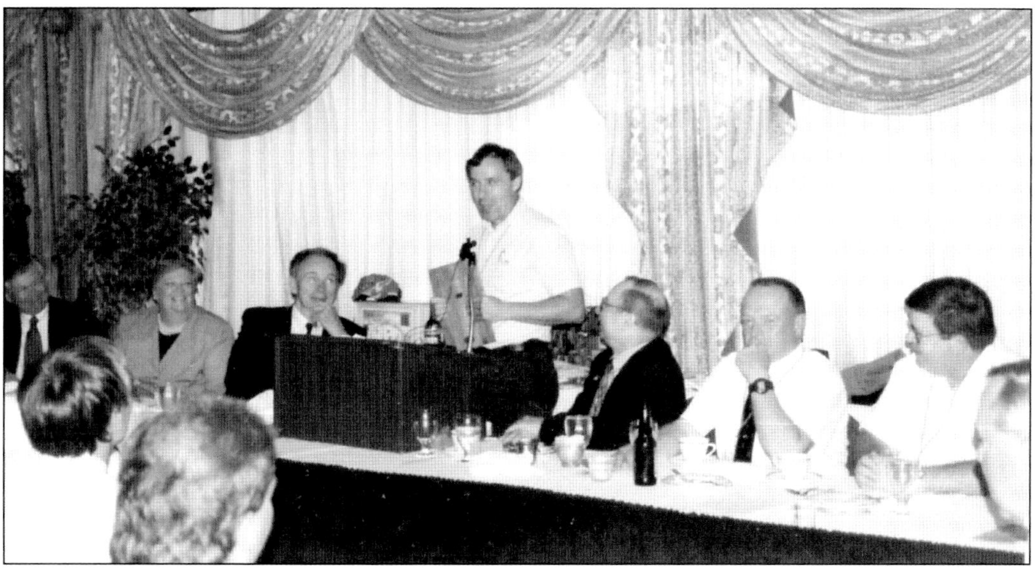

The annual meetings of the NFFS have allowed the 20 states to interact among themselves and with the US Forest Service and have also helped foster camaraderie and pride among the state fire supervisors and meeting participants. At the 35th meeting's banquet, Tom Parent, Maine Forest Service, shares a story after receiving the Northeast Forest Fire Supervisors Award as, from left to right, participants Dan Zimmerman (US Forest Service); Jan Polasky (US Forest Service); George Brooks (US Forest Service); Mike Tirrell, Massachusetts Bureau of Forest Fire Control; Dave Harrison, New Jersey Forest Fire Service; and Coy Mullins, West Virginia Division of Forestry, listen. (New Jersey Forest Fire Service.)

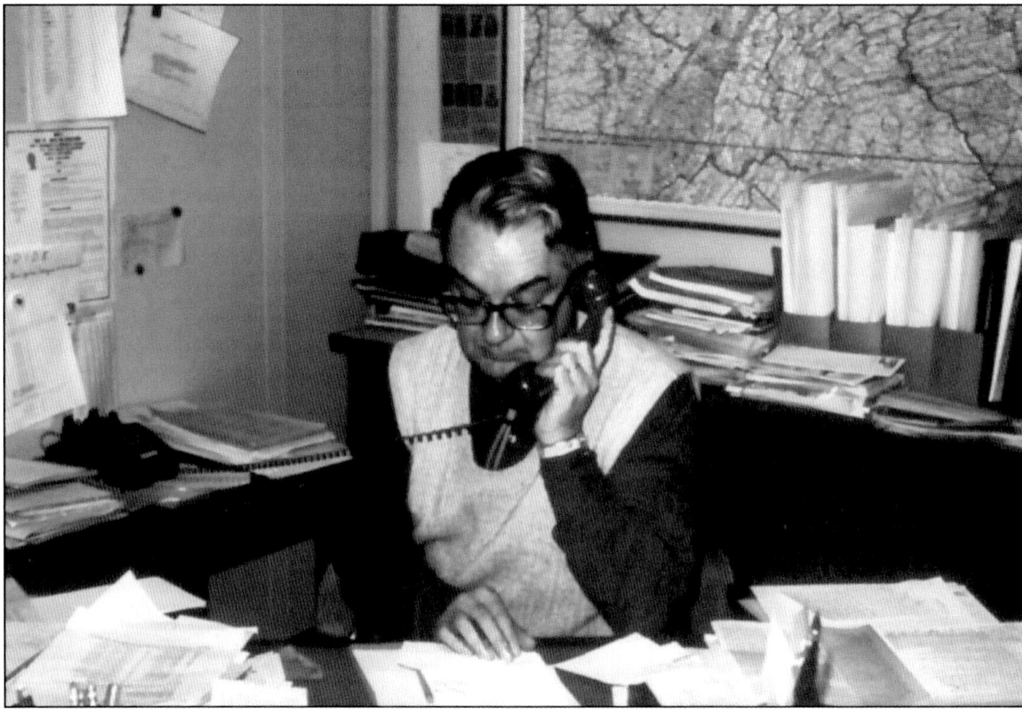

The NFFS Fire Prevention Award is given annually to an individual or organization who has provided outstanding service in wildfire prevention within a state or group of states in the northeastern region. At the 23rd annual meeting of the NFFS, the award was created and named in honor of Eugene F. McNamara, "Mr. Fire Prevention," to honor his life and his commitment to wildfire prevention on a state, regional, and national level. He was Pennsylvania's state fire supervisor from 1966 through 1984. In this photograph, McNamara works diligently at his desk in Harrisburg. (Pennsylvania DCNR Division of Fire Protection.)

McNamara Award Recipients			
1990	Beverly Stout (IN-INS)	2004	Glen Bell (PA-PAS)
1991	Arthur Creelman (PA-PAS)	2005	Great Lakes Forest Fire Compact Prev. Comm.
1992	John Q. Ricard (NH-NHS)	2006	Carol Nillson (MI-HMF)
1993	Alfred E. Grimes (NH-NHS)	2007	Steve Cummings (PA-PAS)
1994	James Downie (ME-MES)	2008	Roxanne Savoie (MA-MAS)
1995	Rose Ann Edmondson (MO-MOS)	2009	Aaron Kloss (OH-OHS)
1996	Philip Stormberg (WI-WIS)	2010	Doug Miner (NH-NHS)
1997	Jim Miller (WI-WIS)	2011	Paul Kollmeyer (MI-MIS)
1998	Paul Sebasovich (PA-PAS)	2012	Amy Luebke (WI-WIS)
1999	Maureen Brooks (MD-MDS)	2013	Rick Meintel (PA-PAS)
2000	Not Awarded	2014	Dave Robbins (MD-MDS)
2001	John Matson (WI-WIS)	2015	Not Awarded
2002	Not Awarded	2016	Fred Turck (VA-VAS)
2003	Catherine Regan (WI-WIS)		

Since 1990, twenty-three individuals and one organization have been awarded the Northeast Forest Fire Supervisors Fire Prevention Award. (NFFS.)

Developing improved techniques in wildland fire prevention has always been a responsibility of the Northeast Forest Fire Supervisors and the 20 states represented. Over the years, more than 35 individuals and groups within the northeastern region have received national Smokey Bear Awards for their dedication to fire prevention. At the 2010 NFFS meeting, Maureen Brooks (US Forest Service) received the Silver Smokey Award for her fire prevention work on the regional level from area fire director Billy Terry (US Forest Service). (Tom Brady.)

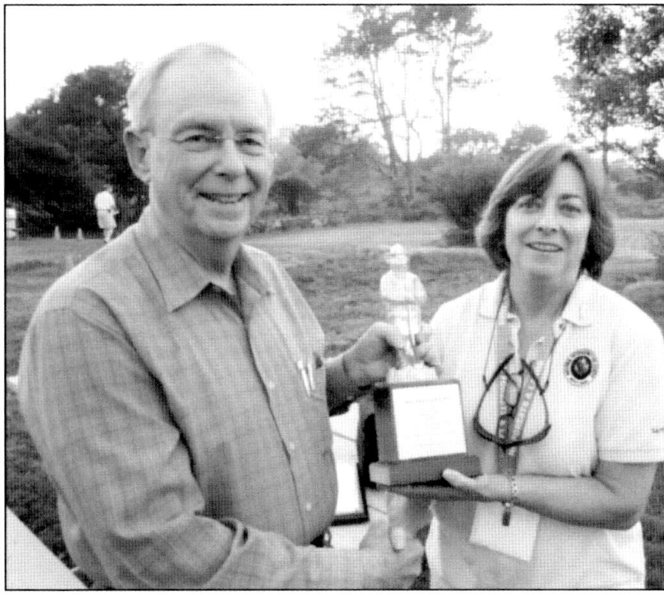

Northeast Forest Fire Supervisors Past Chairs							
1966-67	Mitt Bergman (MI-MIS)	1979-80	Roy Hatcher (IA-IAS)	1992-93	Tom Ronk (MO-MOS)	2006-07	Maris Gabliks (NJ-NJS)
1967-68	Fred Holt (ME-MES)	1980-81	Charles Boone (NY-NYS)	1993-94	Pete Skuba (IL-ILS)	2007-08	Ron Stoffel (MN-MNS)
1968-69	Gene McNamara (PA-PAS)	1981-82	John Kullman (MO-MOS)	1994-95	Nate Kirk (OH-OHS)	2008-09	Brad Simpkins (NH-NHS)
1969-70	Kerwin Hainer (MO-MOS)	1982-83	Jim Cummings (NJ-NJS)	1995-96	Tom Bourn (RI-RIS)	2009-10	Bill Altman (MO-MOS)
1970-71	Max Lane (IL-ILS)	1983-84	Dennis Gardner (MN-MNS)	1996-97	Alan Zentz (MD-MDS)	2010-11	Mike Bowden (OH-OHS)
1971-72	Mal Idleman (OH-OHS)	1984-85	Tim Kaden (DE-DES)	1997-98	Olin Phillips (MN-MNS)	2011-12	Ben Webster (WV-WVS)
1972-73	Bill Phoenix (NJ-NJS)	1985-86	Dave Gillespie (IL-ILS)	1998-00	John Berst (PA-PAS)	2012-13	Monte Mitchell (MD-MDS)
1973-74	Chuck Rieck (WI-WIS)	1986-87	Dave Harrison (NJ-NJS)	2000-01	Michael Terrill (MA-MAS)	2013-14	Andy Jacob (NY-NYS)
1974-75	Asher Kelly (WV-WVS)	1987-88	Steve Creech (IN-INS)	2001-02	Coy Mullins (WV-WVS)	2014-16	Dave Celino (MA-MAS)
1975-76	Bob Compeau (MI-MIS)	1988-89	Joh Bitzer (PA-PAS)	2002-03	Ron Wilson (MI-MIS)	Current	Ron Stoffel (MN-MNS)
1976-77	Oz Herbert (MD-MDS)	1989-90	Jerry Atkins (WV-WVS)	2003-04	Gail Kantak (IA-IAS)		
1977-78	Bill Wilbey (IN-INS)	1990-91	Ed Jacoby (NY-NYS)	2004-05	Trent Marty (WI-WIS)		
1978-79	Jack Sargent (NH-NHS)	1991-92	Bud Nelson (NH-NHS)	2005-06	Bill Williams (ME-MES)		

During the past 50 years, 47 individuals have served as the chairperson of the Northeast Forest Fire Supervisors. They have guided the organization from its beginning in 1966 to today, and it continues to provide key wildland fire management leadership to the 20 northeastern states, as well as the nation. (NFFS.)

Northeast Forest Fire Supervisors Award Recipients			
1968	Horace Rowland (PA-PAS)	1993	Steve Creech (IN-INS)
1969	Leonard McDonald (MN-SUF)	1994	Jack Horton (PA-NAF)
1970	Virgil Hendricks (WXS)	1995	John Wyman (OH-OHS)
1971	Richard Diehl (NH-NHS)	1996	Billy Jack Terry (PA-NAF)
1972	Harry Mitchell (PA-NAF)	1997	Tom Ronk (MO-MOS)
1973	William Martini (WI-WIS)	1998	John Grosman (WI-WIS)
1974	Robert Compeau (MI-MIS)	1999	Alan Zentz (MD-MDS)
1975	William Phoenix (NJ-NJS)	2000	Bud Nelson (NH-NHS)
1976	Fred Holt (ME-MES)	2001	Tom Parent (ME-MES)
1977	Max Bradley (MI-MIS)	2002	John Berst (PA-PAS) Steve Creech (IN-INS)
1978	Max Lane (IL-ILS)	2003	Ron Wilson (MI-MIS)
1979	Dick Mullavey (PA-NAF) Duane Dupor (WI-WIS)	2004	Gail Kantak (IA-IAS) Tom Tolska (NJ-NJS)
1980	Gene McNamara (PA-PAS)	2005	Don Artley (NASF)
1981	Don Grant (MI-MIS)	2006	Maris Gabliks (NJ-NJS)
1982	Gordon Landphier (WI-WIS)	2007	Matt Dillon (WV-WVS) Mike Folgert (WI-WIS)
1983	Bob Stowell (MI-MIS)	2008	Charlie Keller (IN-INS)
1984	Bill Herbolsheimer (PA-NAF)	2009	Trent Marty (WI-WIS)
1985	Jim Miller (WI-WIS)	2010	Kirk Bradley (MI-MIS)
1986	Jack Sargent (NH-NHS)	2011	Brad Simkins (NH-NHS)
1987	John Kullman (MO-MOS)	2012	Jan Polasky (PA-NAF)
1988	Amy Adams (PA-NAF)	2013	Mike Bowden (OH-OHS)
1989	Dave Harrison (NJ-NJS)	2014	Scott Heather (MI-MIS)
1990	Dennis Gardner (MN-MNS)	2015	Bob Hartlove (PA-NAF)
1991	Joe Hughes (NJ-NJS)	2016	Tom Brady (PA-NAF)
1992	Jacqueline Allen (PA-NAF)		

Since 1968, fifty-two individuals have received the annual Northeast Forest Fire Supervisors Award. One individual, Steve Creech, has received the award twice. In addition, Jim Barresi, Mac Gramley, and Eric Jager received Northeast Forest Fire Supervisors' Special Recognition Awards for their work in support of the northeastern states' forest fire management agencies. (NFFS.)

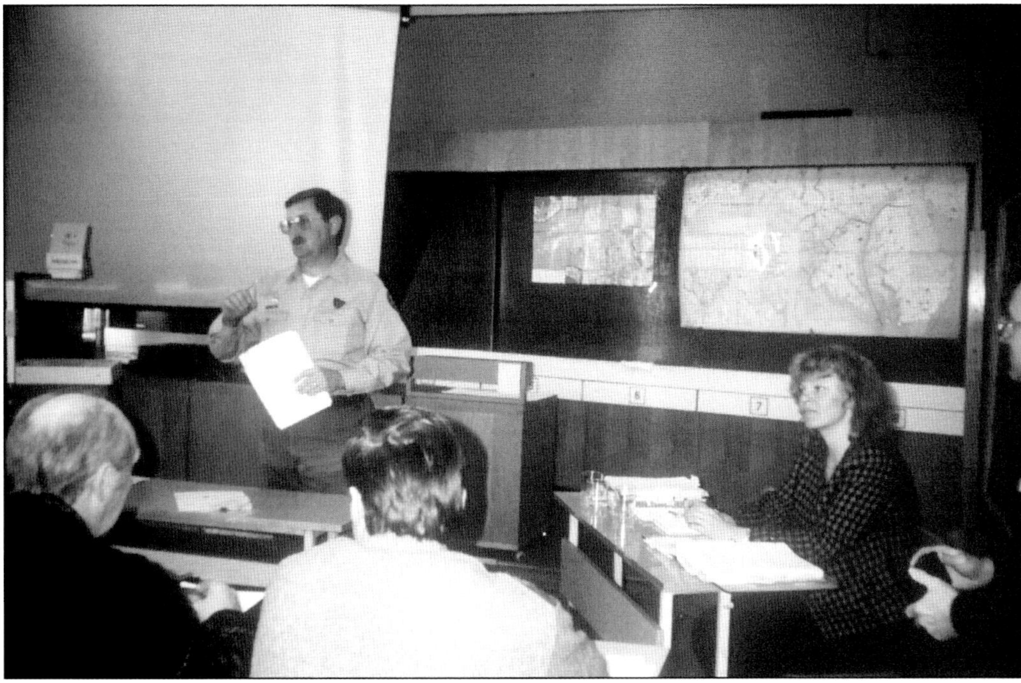

The NFFS Award is given annually to the person who has made significant contributions in the field of forest fire protection from the geographic area making up the northeastern area. At the 44th annual meeting of the NFFS, the award was renamed the H. Alan Zentz Fire Supervisors Award to honor his life and his commitment and contributions to forest fire protection throughout his 35-year career with both the Maryland and the US Forest Services. In this photograph, Zentz instructs a training course for the Maryland Forest Service. (NFFS.)

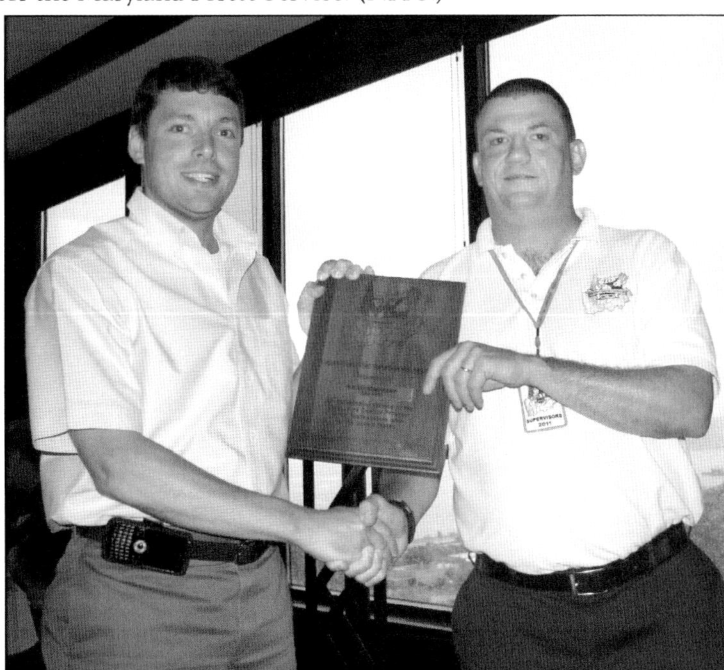

At the 45th NFFS meeting, Brad Simpkins (left), of New Hampshire Division of Forests and Lands, received the annual NFFS Supervisors Award, presented by chairperson Mike Bowden (right), of Ohio DNR Division of Forestry. This was the inaugural year for the NFFS Award being named in honor of H. Alan Zentz. Zentz received the award in 1999, was a past chair of the organization, and had a 35-year career with the Maryland and the US Forest Services. (NFFS.)

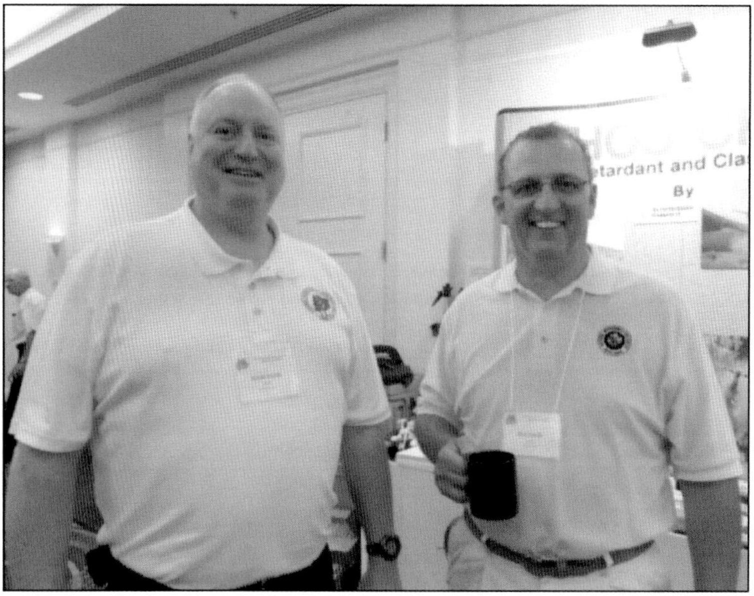

The Northeast Forest Fire Supervisors has worked very closely with the National Association of State Foresters (NASF) Forest Fire Protection Committee in addressing state wildland fire protection issues that impact the nation on an ongoing basis. Tech support director Keith Smith (left) and NASF fire director Dan Smith (right) meet with fire supervisors during an annual meeting. (Tom Brady.)

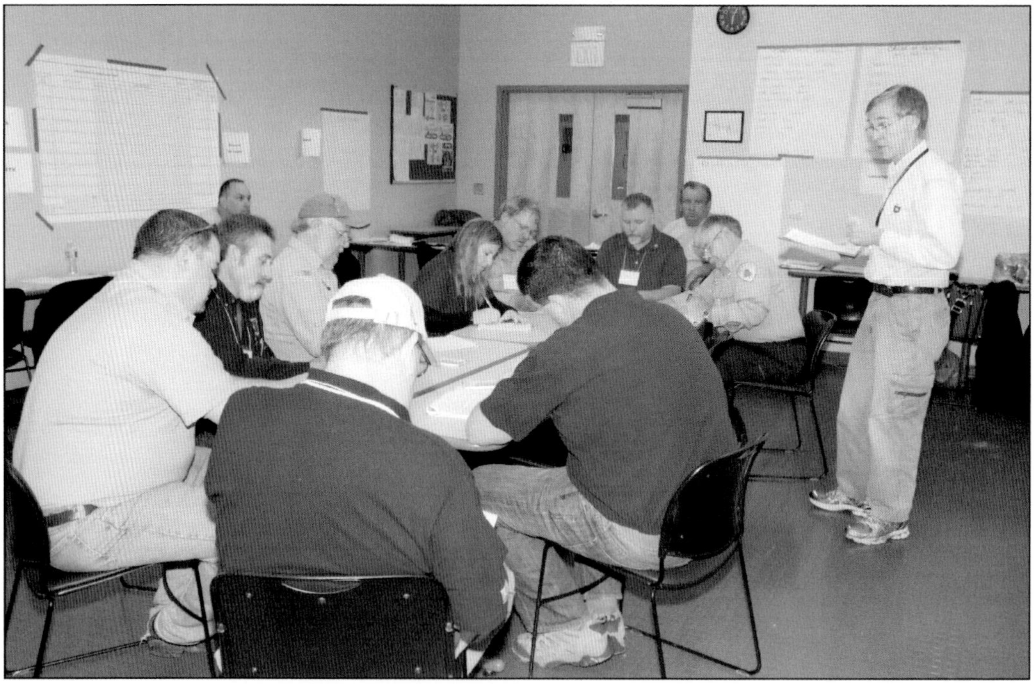

John Berst (MAIFFPC administrator and retired Pennsylvania DCNR Division of Fire Protection fire supervisor) briefs Incident Management Team (IMT) students during a S420 Command and General Staff course hands-on training simulation. The NFFS have been leaders in providing incident management and leadership training to their 20 member states. (Ken Badger.)

Two

Northeastern States

The northeastern states are comprised of Maine, New Hampshire, Vermont, Rhode Island, Connecticut, Massachusetts, and New York. The concept of creating a forest fire compact to provide mutual aid between states in combatting wildfires came after the disastrous 1947 fires in Maine that destroyed nine communities and burned over 250,000 acres of forest. After the passage of authorizing law by the Congress of the United States and by confirming legislation in each state, the first forest fire compact was established in 1949 and became the Northeastern Forest Fire Protection Compact (NFFPC). Realizing that forest fires could cross international borders, Congress authorized Canadian provinces to join the compact in 1952. When Quebec joined in 1969 and then New Brunswick in 1970, the compact became the first international wildland fire entity. In 1996, Nova Scotia became a full member, and the Green and White Mountain National Forests became associate members. The province of Newfoundland and Labrador (one province) joined the compact in 2007. The US Fish and Wildlife Service and the National Park Service became associate members in 2011, as did the Fire Department of New York City in 2015. Finally, the province of Prince Edward Island became a full member in 2016, resulting in 12 full members and 4 associate members.

The compact has a governing body called the Northeastern Forest Fire Protection Commission. Each state is entitled three commissioners represented by their state forester, a state legislator, and a representative of their governor. Canadian provinces also provide three commissioners, while associate members provide two commissioners and are nonvoting. The US Forest Service's Northeastern Area State & Private Forestry provides assistance to the compact on an annual basis. Since the compact's inception, it has formed committees and working teams to develop and coordinate plans, to provide standardized training, and to improve and coordinate forest fire protection across the geographic area. Currently, the compact has working groups and committees with representation from every member agency that addresses firefighting operations, training, prevention and education, equipment and technology, fire science, forest health, and resource sharing.

The states within the Northeastern Forest Fire Protection Compact range from the pine barrens on Long Island to New England hardwoods and large mature softwood tracts covering over 225 million acres of protected forestland. Each state has its own significant wildland fire protection challenges and has developed diverse state fire protection programs to deal with their situations.

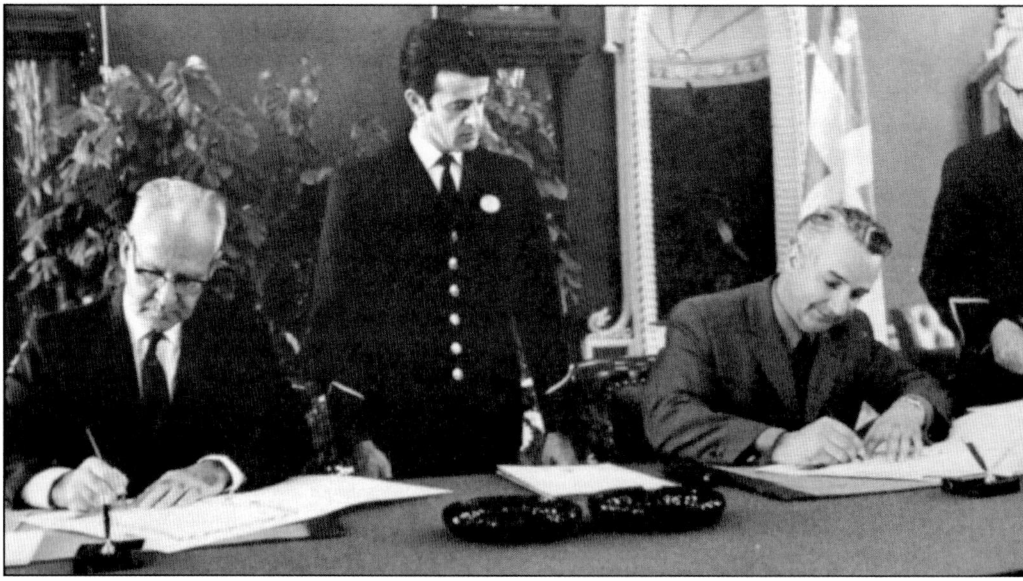

At the far left, Austin Wilkins, former commissioner of the Department of Forestry for the State of Maine, is signing the NFFPC agreement in Quebec on September 23, 1969. Major fires of 1947 in Maine leveled nine towns and required Coast Guard evacuations from Bar Harbor, Maine, on Mount Desert Island. These fires were the impetus for establishing NFFPC as the first forest fire protection compact in the United States in 1949. Quebec was the first Canadian province to join a compact in the United States, thereby making the organization international. Wilkins was a key player in initiating and pursuing formation of the compact. He lived to be 101 years old. (NFFPC.)

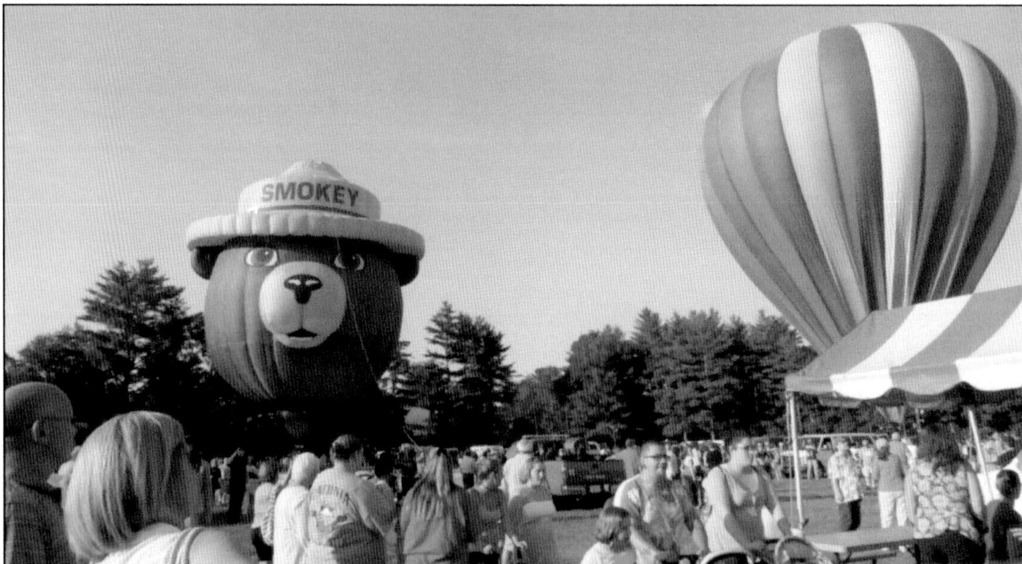

In 2014, a large crowd attended the annual balloon festival in Hillsborough, New Hampshire. The Smokey Bear balloon was the highlight of the event. Festival participants could take rides in the balloon throughout the day. The northeastern compact's Fire Prevention Working Team planned, coordinated, and managed all of the Smokey Bear rides and fire prevention events at the festival. (NFFPC.)

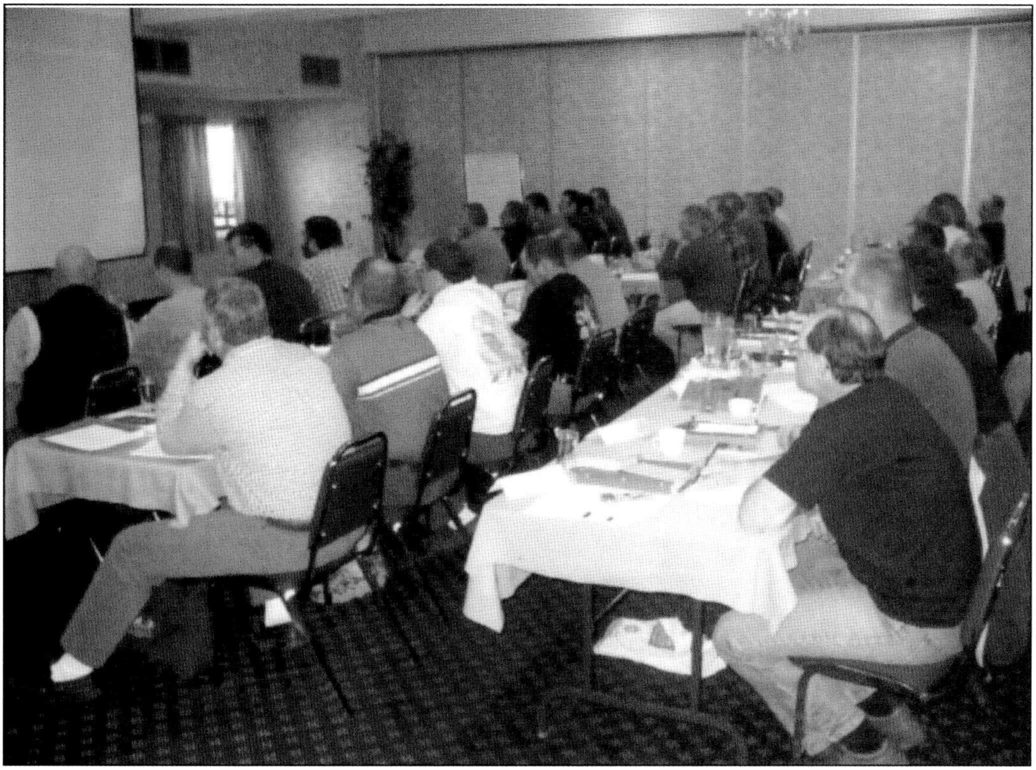

The northeastern compact conducts as many as four training academies per year in various locations intended to best meet the needs of their members. This photograph is of a fire academy in Portland, Maine, providing nationally certified courses to develop Incident Management Teams. The compact has trained many Incident Management Teams in preparation to meet agency needs and to share resources throughout the United States and Canada. (NFFPC.)

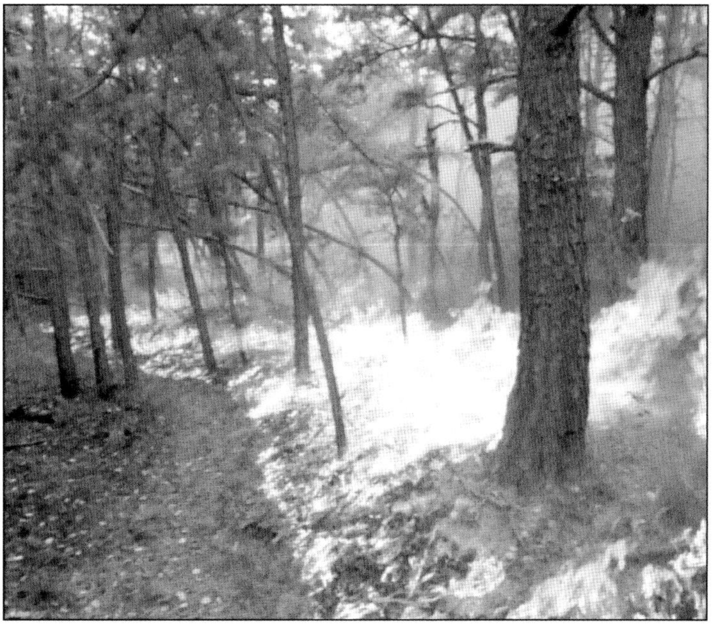

Prescribed burning is an important part of any fire management program. The Fire Science Working Team of the northeastern compact works to enhance collaboration on projects such as this. The team addresses many other aspects of fire science and offers fire behavior training, weather forecasting via the National Weather Service and shared compact-owned remote weather stations, and promoting and supporting fire research for the area. (NFFPC.)

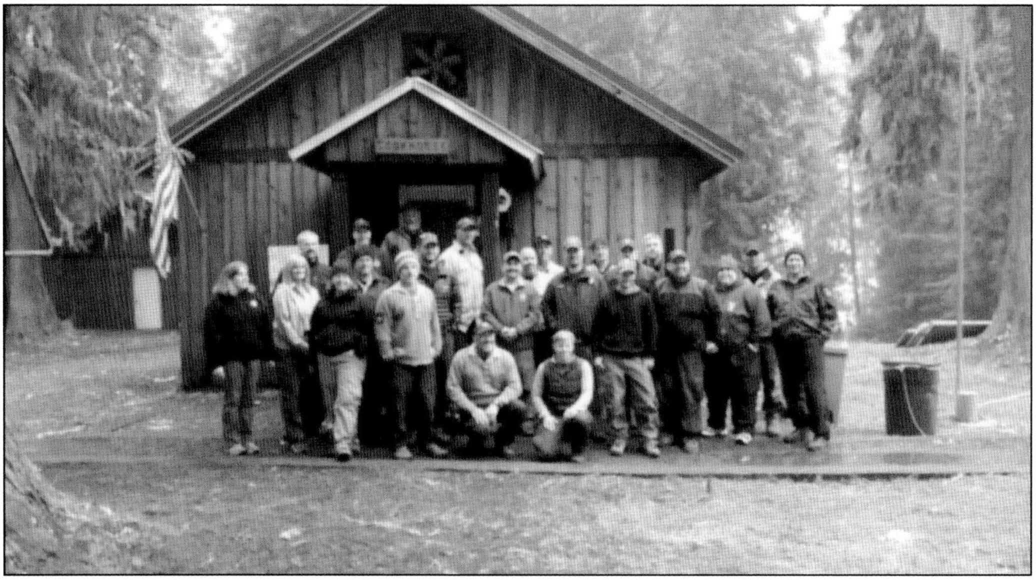

In 2015, a Northeastern Forest Fire Protection Compact IMT was mobilized to manage the Last Inch fire in Idaho. The IMT was dispatched as a compact-to-compact resource exchange. Team members were from NFFPC states, the province of New Brunswick, and member federal agencies from the compact. (NFFPC.)

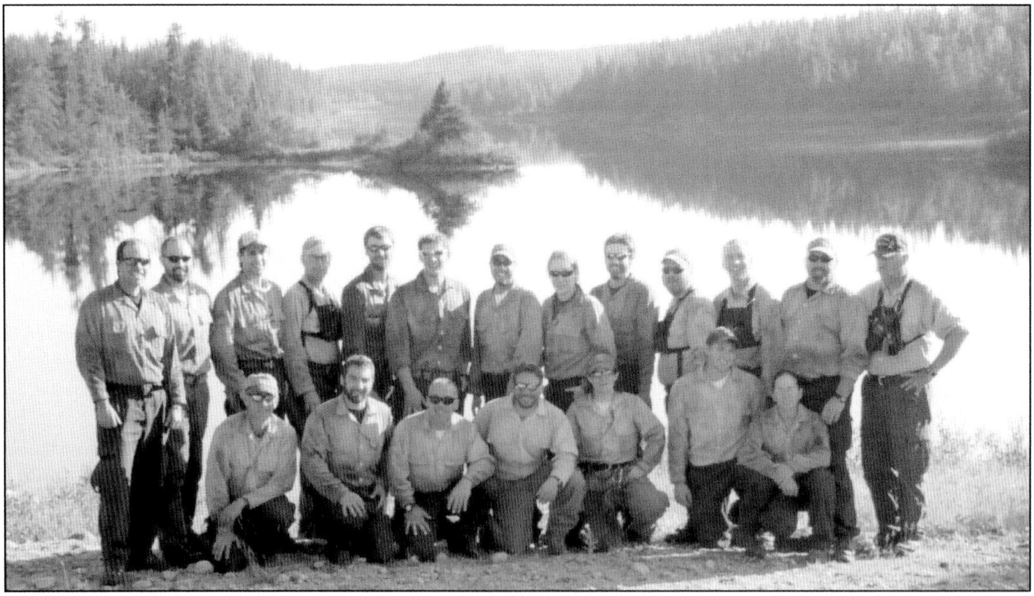

Sharing resources within the northeastern compact is commonplace, and in 2015, the compact successfully mobilized a Quebec crew to a wildfire in Idaho as a compact-to-compact resource exchange. This was the first time that the northeastern compact sent a Canadian crew to help a state from another compact jurisdiction. This photograph shows the Quebec crew after being briefed for their assignment. (NFFPC.)

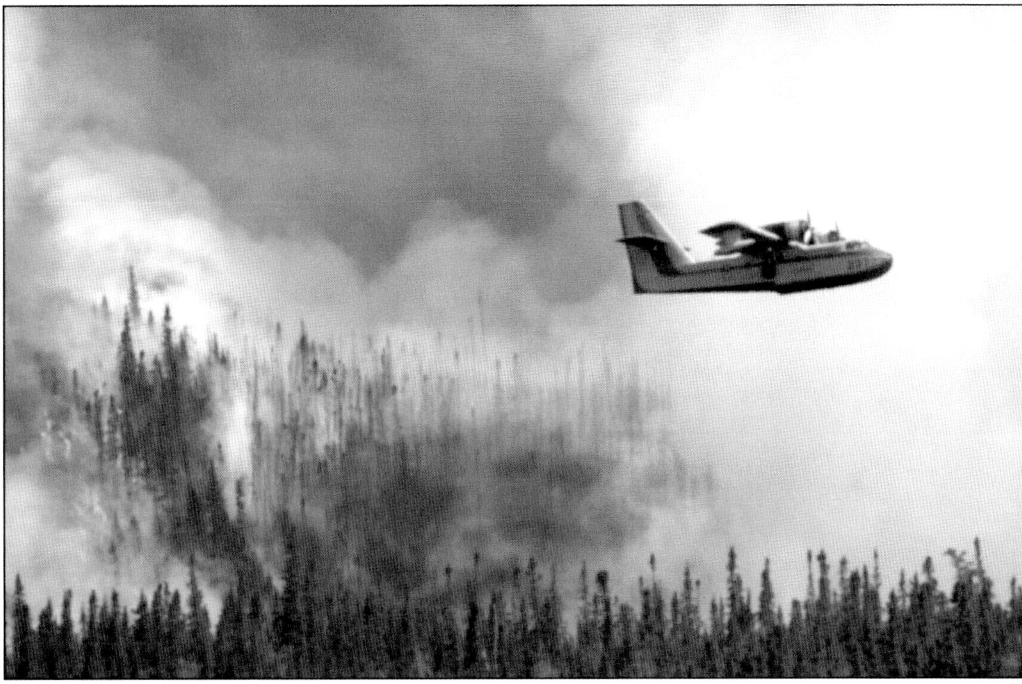

The NFFPC's ability to share aviation assets among compact members, including internationally, is a major benefit of the compact organization in combating wildfires across the United States and Canada. This photograph captures a CL-415 air tanker from Quebec working a fire. (NFFPC.)

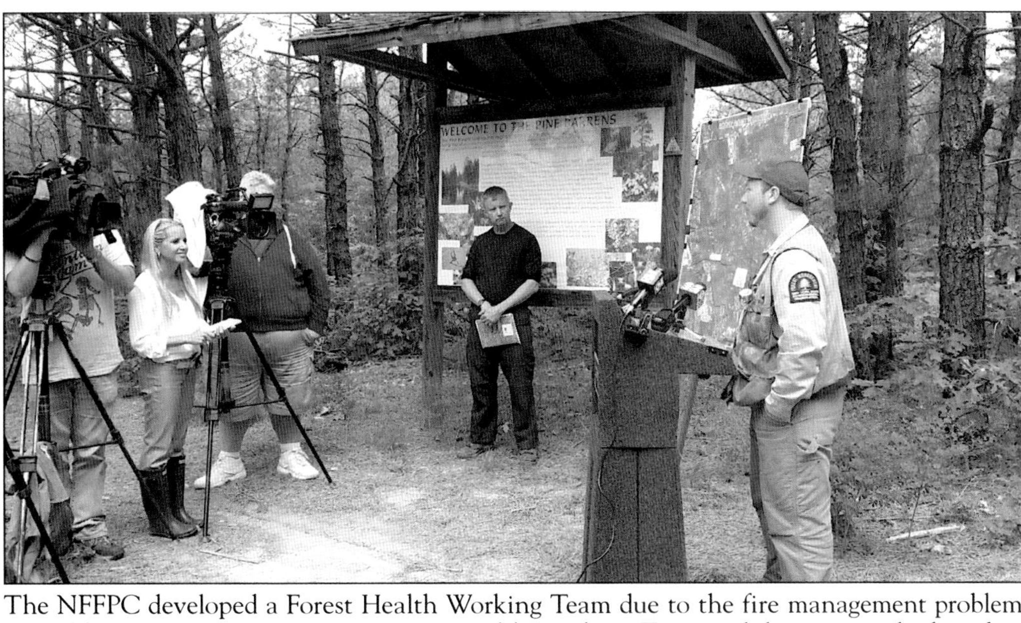

The NFFPC developed a Forest Health Working Team due to the fire management problems caused by insects. Seven separate Forest Health Working Team mobilizations took place from 2013 through 2015 to address emerald ash borer, Asian long-horned beetle, and Southern pine beetle problems within the compact area. The use of the Incident Command System to manage these incidents was deemed valuable to Forest Health Working Team operations. This photograph shows a team working in Long Island. (NFFPC.)

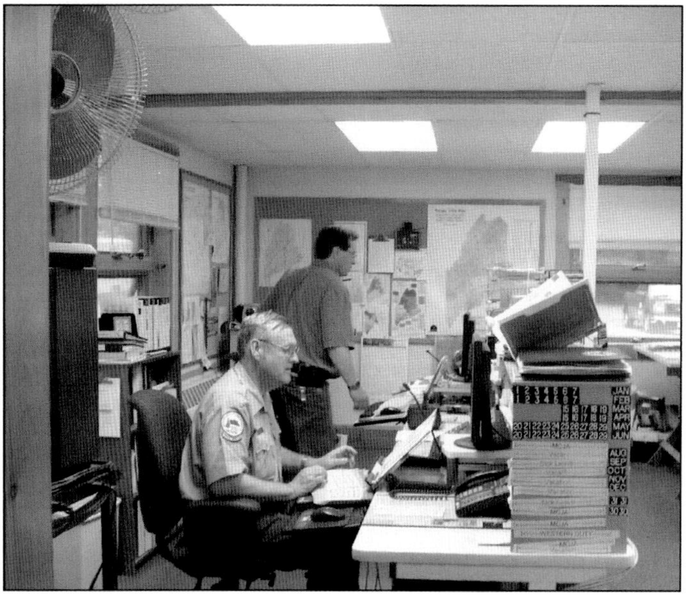

This photograph shows dispatchers working at the northeastern compact's dispatch center in Augusta, Maine, which was established in 2001. It includes a cost-sharing agreement among the compact and federal partners. It is a unique center that mobilizes resources within the United States, the compact area, and between compacts within the United States and Canada. The center was moved to the White Mountain National Forest office in 2015 but still operates under the same agreement and structure. (NFFPC.)

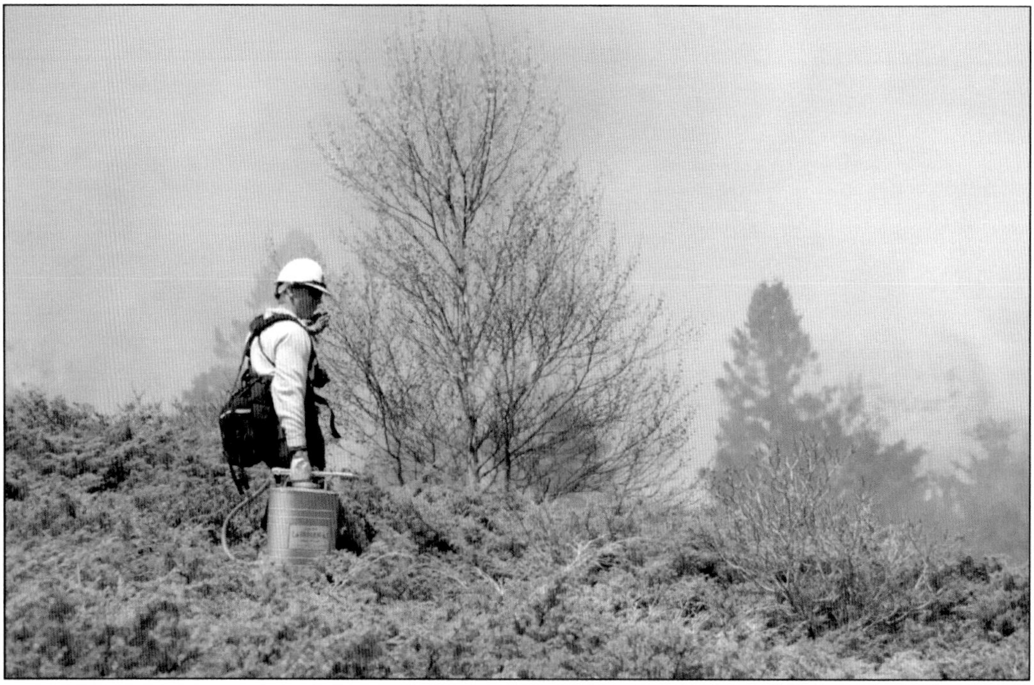

In this photograph, a Maine forest ranger sizes up a wildfire burning on an island. Over 200,000 acres of forest were burned, and hundreds of homes were destroyed by wildfires in October 1947. After the 1947 wildfires, the Maine Forest Service became responsible for protecting the entire state from wildfires, which encompassed 450 municipalities. (Maine Forest Service.)

This 1954 photograph shows the Maine Kineo Mountain Forestry District station. Maine is the most heavily forested state in the nation with 17.7 million acres of forests. The position of Maine forest commissioner was created in 1891, and in 1909, the Maine Forestry District was created to protect 10 million acres of forest from wildfire in the state's unorganized territories. (Maine Forest Service.)

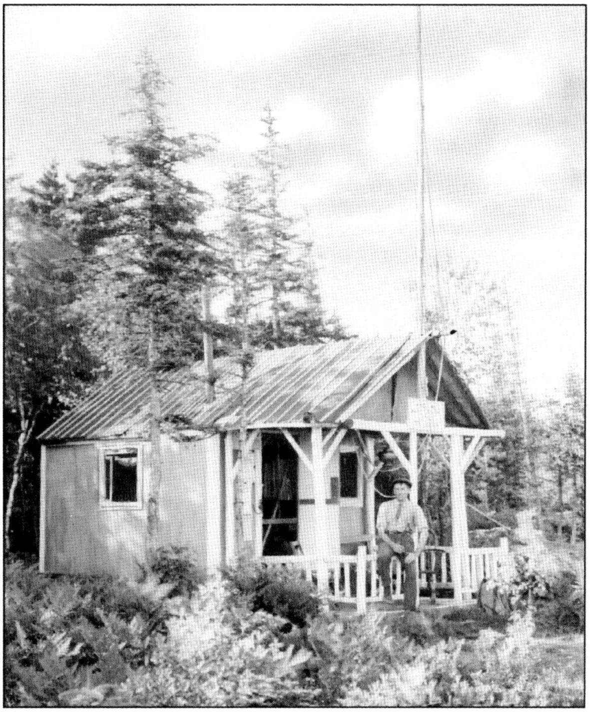

Maine was the first state east of the Mississippi River to utilize aircraft for detecting wildfires. This photograph captures a Cessna 180 loading up with firefighting supplies at Greenville in 1958. Maine is known as the "Vacationland" state as noted on the fire patrol vehicle's state license plate. (Maine Forest Service.)

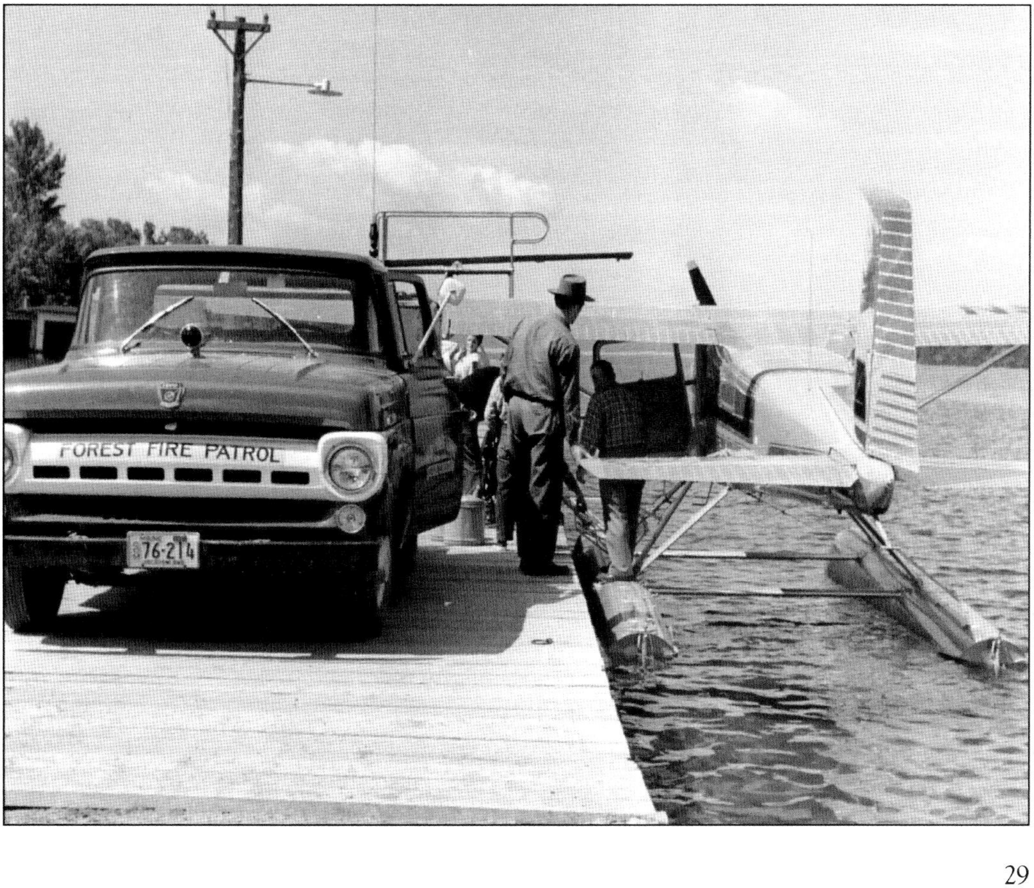

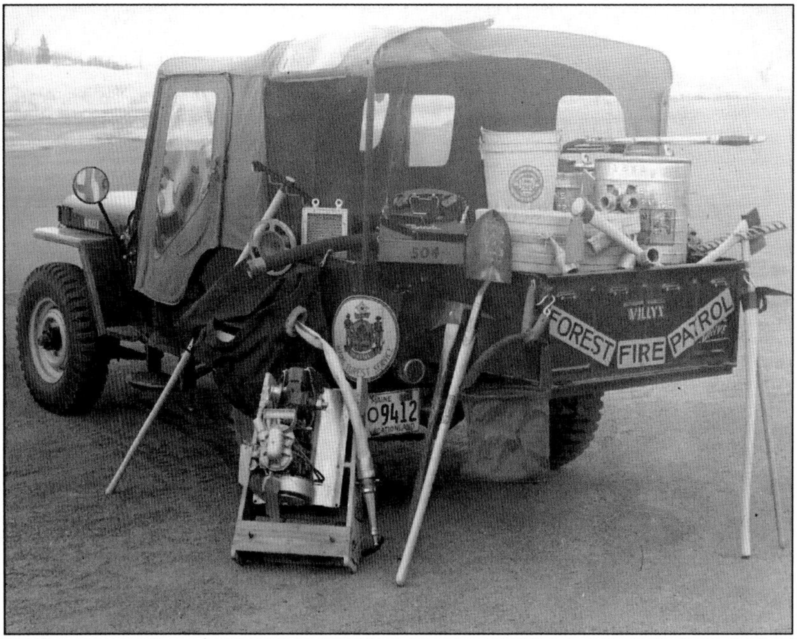

A Maine Forest Service "Forest Fire Patrol" Willy's Jeep was photographed in 1951. The jeep contains firefighting hand tools and equipment from that era, which included hose, nozzles, shovels, Pulaski tools, various rakes, portable pump, buckets, portable radio, and backpack pumps. (Maine Forest Service.)

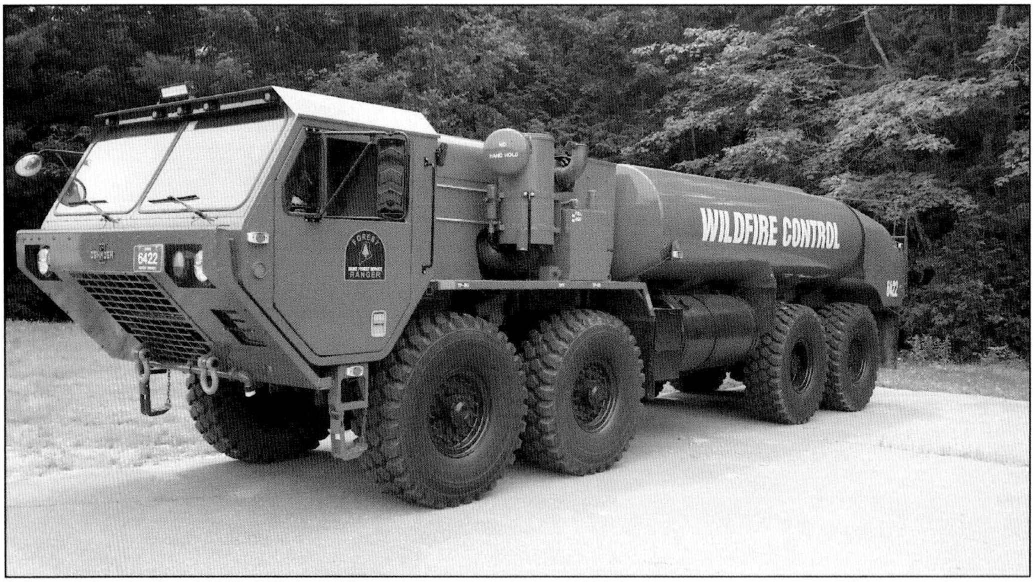

This Heavy Expanded Mobility Tactical Truck (HEMTT) was acquired by the Maine Forest Service through the Department of Defense's Firefighter Property program and developed into an off-road engine. It is an eight-wheel drive and carries over 2,000 gallons of water. This type of vehicle has been manufactured for the military since 1982, and over 27,000 have been manufactured as of 2015. (Maine Forest Service.)

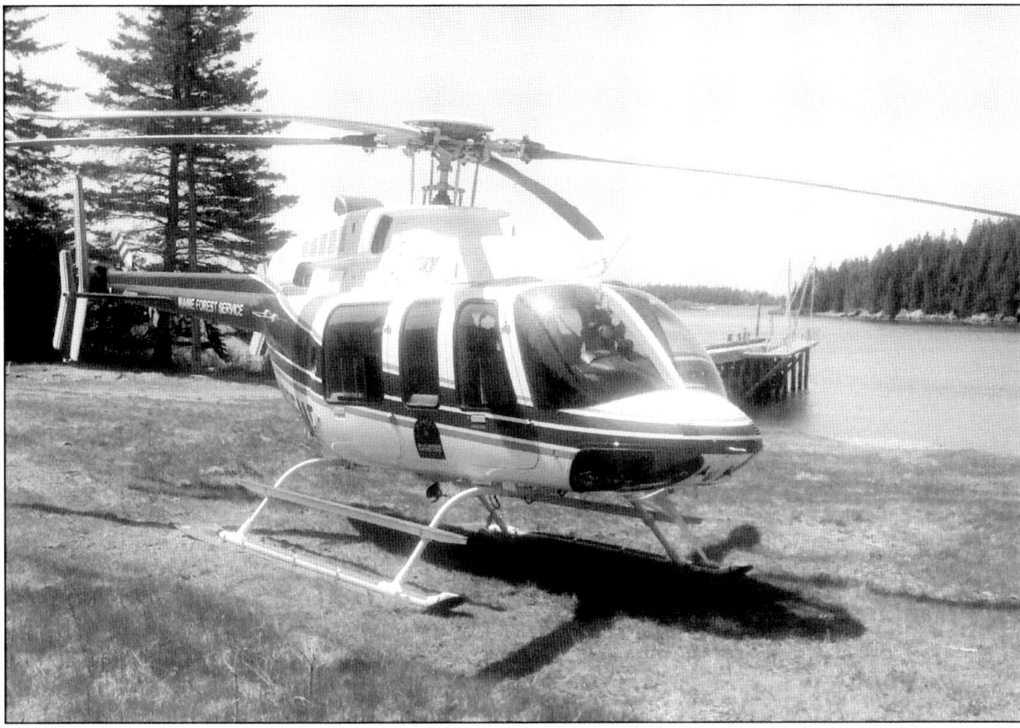

A Bell 407 Type 3 helicopter sits at a helispot near one of Maine's thousands of lakes. The Maine Forest Service has utilized helicopters since the 1950s to help with their forest protection mission. Today, their aviation fleet consists of a mix of Bell 407 and Bell Huey helicopters. (Maine Forest Service.)

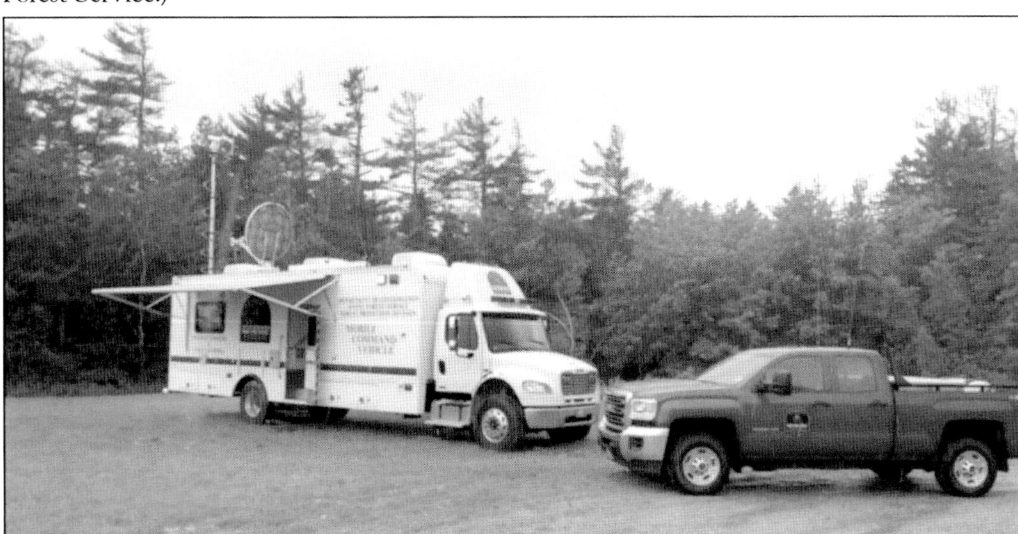

This 2015 photograph shows a current Maine forest ranger patrol vehicle, as well as their mobile command vehicle. The Maine Forest Service operates one of the mobile command vehicles out of a statewide fleet of four. It serves as mobile command to control and to assist with the management and communication of complex wildfires and all-hazard emergency incident responses. (Maine Forest Service.)

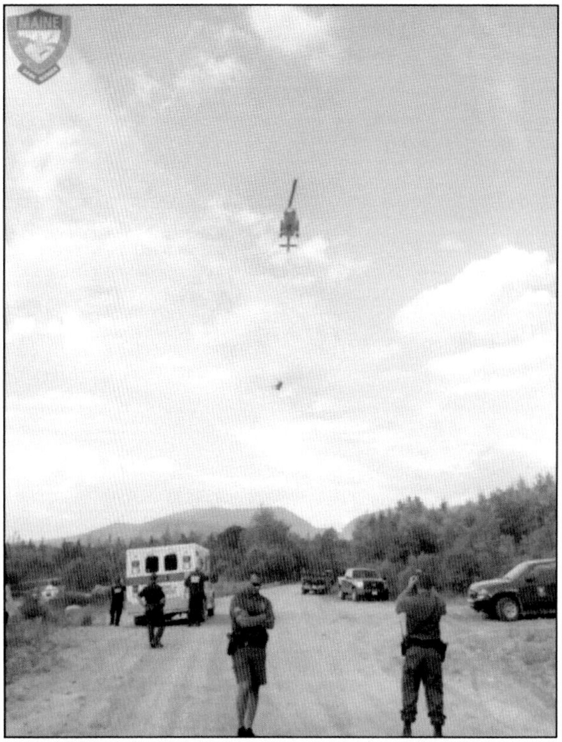

A Maine Forest Service helicopter is seen using the short-haul rescue technique to quickly and safely extract a hiker from Georgia who was injured while hiking the Appalachian Trail through Maine near the New Hampshire border in 2016. This rescue method avoided an 8-to-10-hour carry out of the injured party by using a 12-minute flight helicopter extraction. Maine forest rangers also utilize the helicopter for inserting firefighters into remote wildfires, as well as delivering firefighting gear, supplies, and cargo. (Maine Forest Service.)

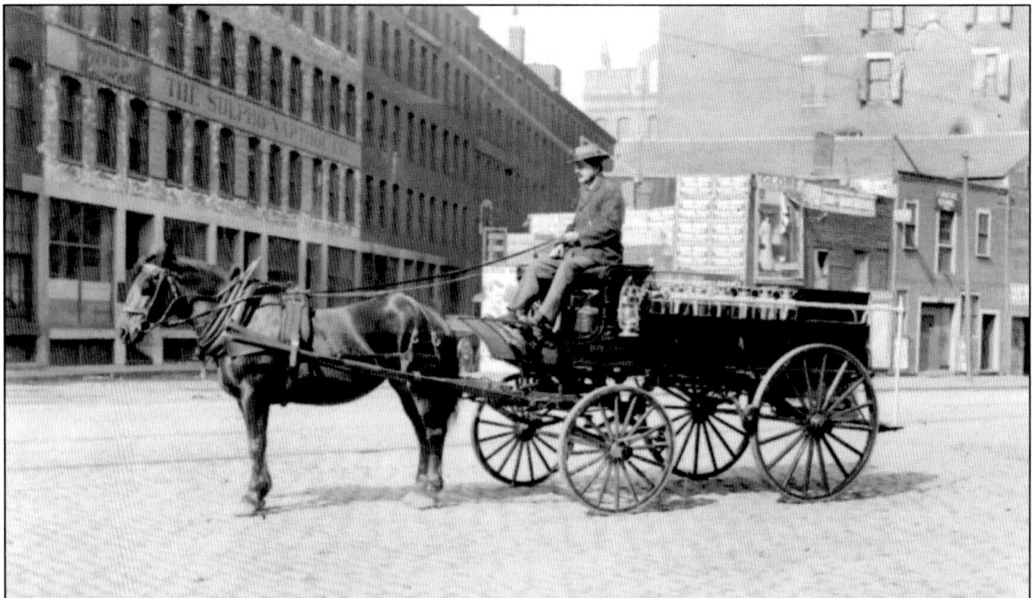

In 1909, Frank W. Rane, the first Massachusetts State forester, built and outfitted two state-of-the-art fire wagons. One was priced at $300 and a second at $450. Each wagon carried four-to-eight chemical extinguishers, galvanized cans containing two extra charges of water, shovels, rakes, mattocks, and spare chemical charges. Forester Rane toured the commonwealth with the units, so city and town officials could see examples of the current wildland firefighting equipment. (Massachusetts Bureau of Forest Fire Control.)

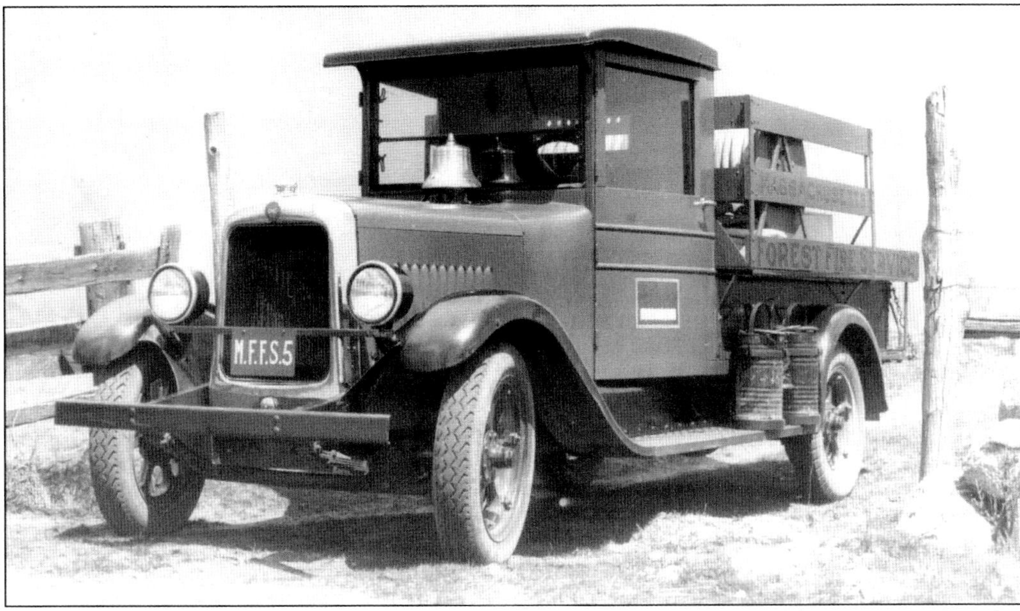

Massachusetts purchased its first state wildland fire engine in 1916, and by 1920, one hundred of the 365 Massachusetts communities had automobiles ready to combat wildfires. In this 1927 photograph, district fire warden Charles L. Woodman operated this GMC during the 1920s. It carried a steel water tank, a gasoline-powered Fitzhenry Guptil water pump, forestry hose, assorted hand tools, and portable water extinguishers. (Massachusetts Bureau of Forest Fire Control.)

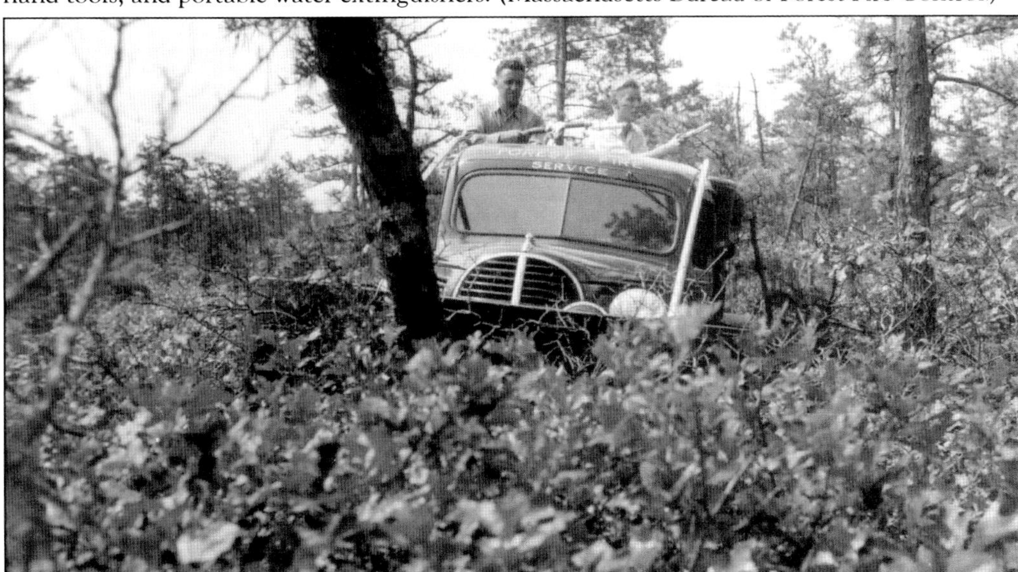

By the 1930s, brushbreakers, fire engines unique to southeast Massachusetts, had been developed. The area's sandy soil, which is not conducive to deep-rooted vegetation, makes it easy to push over small trees with the right equipment. Using a powerful motor, with reinforced bumper and bar work, the breaker driver will carefully push through selective paths to create a containment line. Utilizing plumbing attached to the body enables a wet line to be added as needed. Hose operators complement this by directing water to critical areas. (Massachusetts Bureau of Forest Fire Control.)

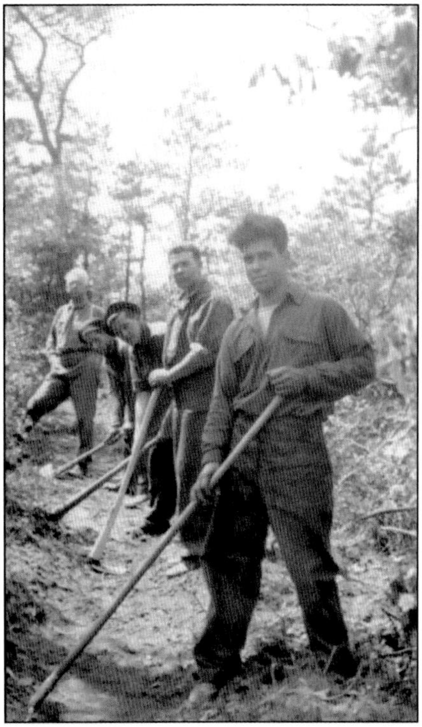

With many Americans out of work in the 1930s, the federal government created the Civilian Conservation Corps (CCC). Massachusetts's CCC crews, like this one photographed in 1937, devoted considerable labor to hazard reduction, firefighter access, and park improvements. During fire season, trained CCC crews of 20 to 30 young men were available to assist with fire suppression. Each CCC camp was equipped with one or more portable pumps, over 2,000 feet of 1.5-inch hose, a complement of hand tools, and 30 backpack fire pumps. (Massachusetts Bureau of Forest Fire Control.)

Beginning in the 1950s, the State of Massachusetts converted its fleet of wildland fire engines to the Dodge Power Wagon. The first Power Wagon was built for the military during World War II. The civilian model was advertised as "the truck that needs no roads," and in the fire service, this was proven to be fact. This 1958 photograph captures a four-by-four Power Wagon outfitted with a PTO winch, 200-gallon steel water tank, 3,000 feet of forestry hose, assorted hand tools, water cans, and mobile radio. (Massachusetts Bureau of Forest Fire Control.)

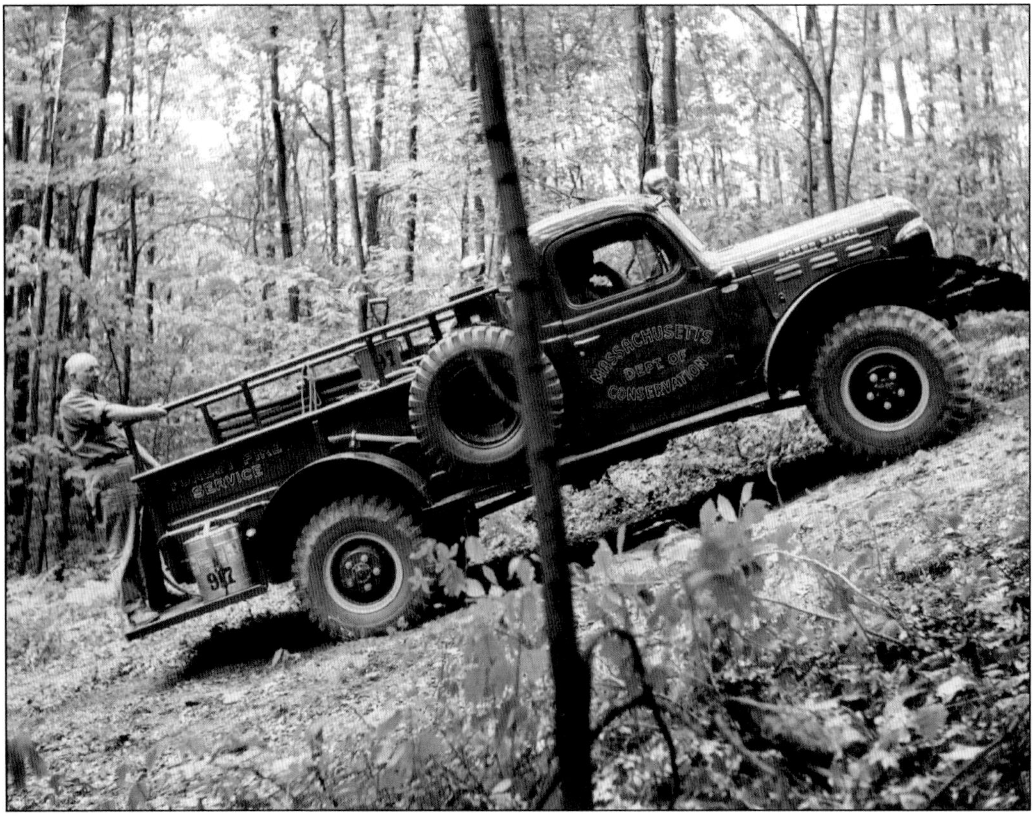

Preventing wildfire was apparent as early as 1907 when Massachusetts posted fire laws and fire bans at railroad stations, post offices, and other areas of focus to inform the public. In 1910, pamphlets, titled *We Must Stop Forest Fires in Massachusetts,* were created and distributed. In the 1930s, motorists were welcomed at fire towers to promote both prevention work and the outdoors. By the 1940s, fire personnel visited schools to promote safety to children, and since the 1950s, Smokey Bear has become an icon, as depicted in this 1969 photograph. (Massachusetts Bureau of Forest Fire Control.)

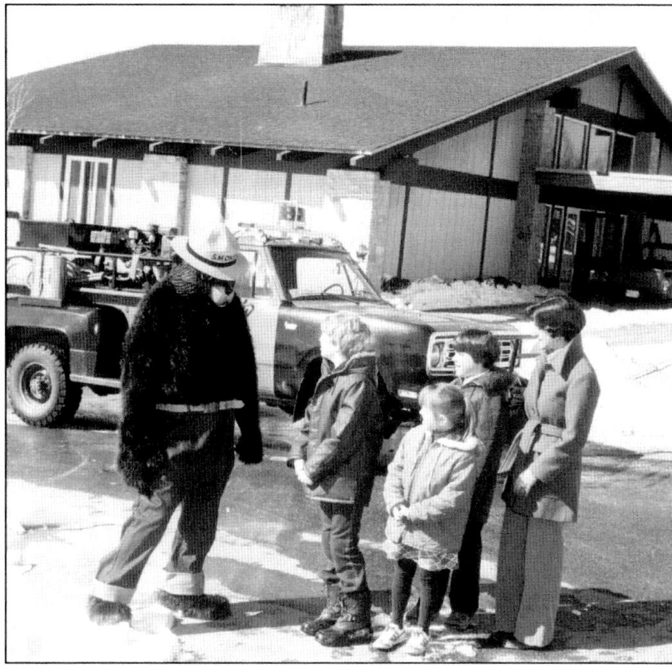

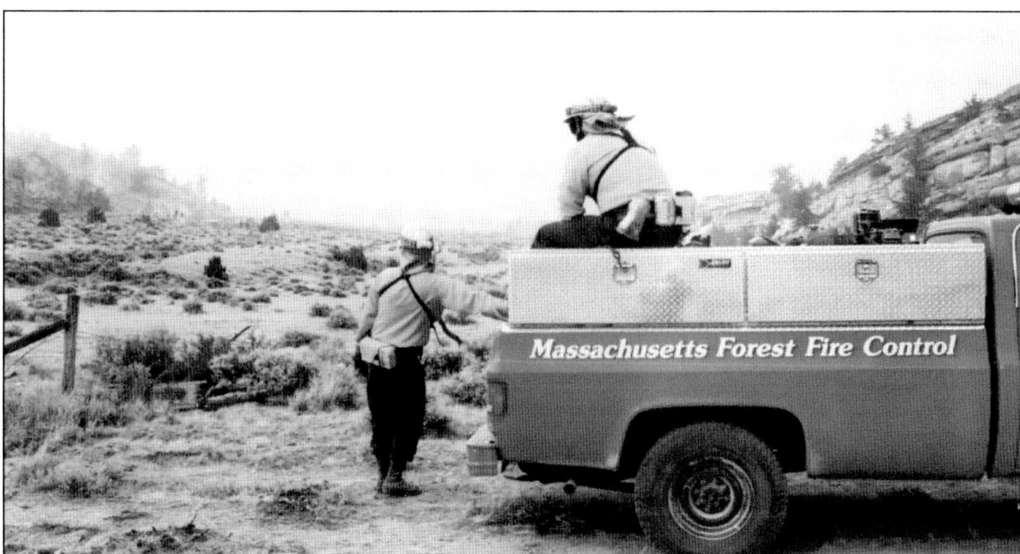

In 2000, a Massachusetts engine strike team with 11 firefighters was dispatched to the Kate's Basin Complex outside Riverton, Wyoming. Massachusetts has long responded to calls for assistance from out of state. In 1921, New Hampshire requested aid extinguishing a fire burning over Mount Madison in Randolph. In 1949, Massachusetts joined with other New England states to form the Northeast Forest Fire Protection Commission, and in 1992, the first handcrew was mobilized west through agreement with the US Forest Service. (Massachusetts Bureau of Forest Fire Control.)

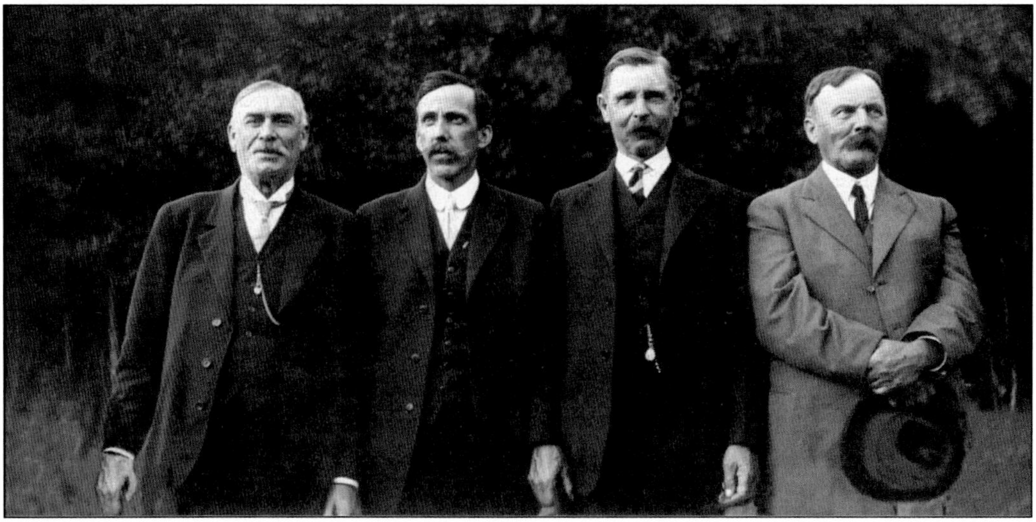

These four men comprised the State of New Hampshire Forestry Department's Forest Fire Service in 1916. Pictured from left to right are district chiefs John J. McNulty, W.H. Langmaid, W.H. Morrison, and Frank P. Allard. New Hampshire's first state forester was appointed in 1909, and that same year, the state's first fire prevention laws were created. (New Hampshire Division of Forests and Lands.)

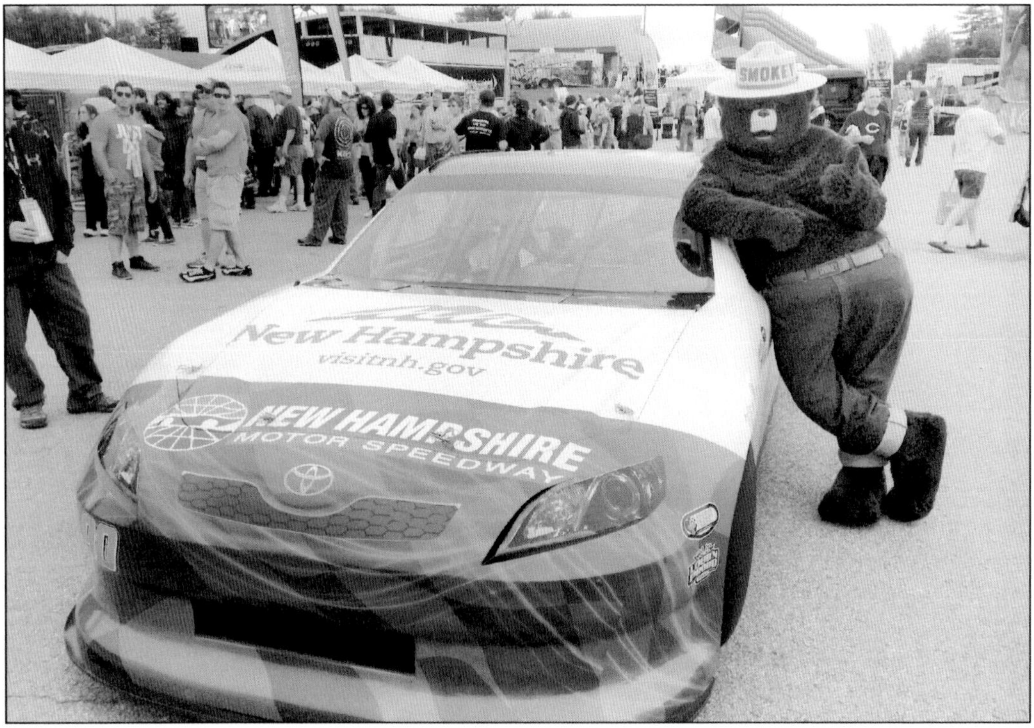

Here is Smokey Bear at the NASCAR race at the New Hampshire Motor Speedway in 2014. Over the years, New Hampshire forest rangers have been bringing Smokey to the NASCAR races to meet with thousands of fans and spread his fire prevention message. This is just one of many high-profile events where the New Hampshire Division of Forests and Lands is able to promote Smokey Bear and showcase fire prevention. (New Hampshire Division of Forests and Lands.)

Here, Massachusetts wildland firefighters build a line at the 2010 Tekoa Mountain wildfire that occurred from April 11 to 15 and burned over 321 acres in the towns of Montgomery and Russell. Firefighters from 14 departments, the Army Air National Guard, Massachusetts State Police Air Wing, American Red Cross, and Massachusetts Department of Fire Services contributed to the effort. Firefighting resources included engines, pumps, helicopters with water buckets, ATVs, and handcrews using rakes, shovels, Pulaskis (axes), and leaf blowers. (Massachusetts Bureau of Forest Fire Control.)

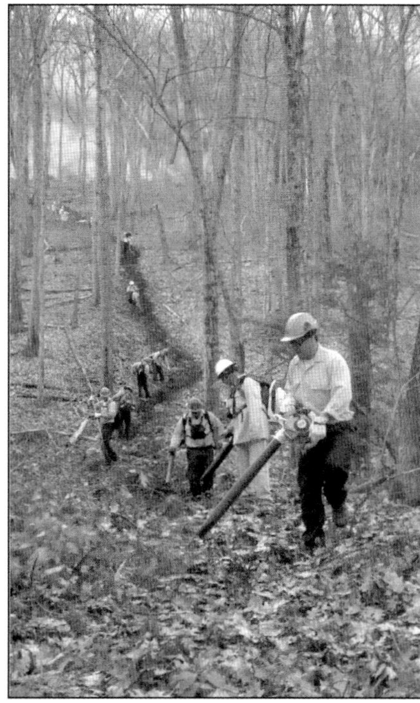

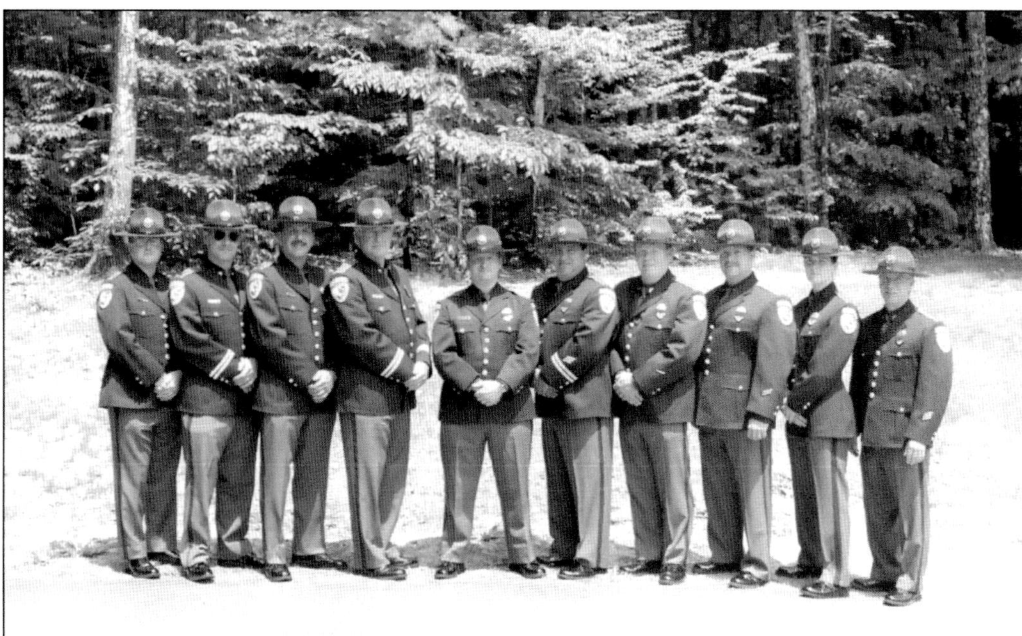

New Hampshire forest rangers in dress uniforms attended the funeral of retired forest ranger Alfred E. Grimes in 2008. Grimes and his wife, Sylvia, were recipients of the Silver Smokey Bear Award in 1994 and the Golden Smokey Bear Award in 1997 for their dedication to fire prevention. Pictured are, from left to right, Neil Bilodeau, John Dodge, John Accardi, Bert VonDohrmann, Steven Sherman, Bryan Nowell, Douglas Miner, Robert Boyd, Jamie Sheehy, and Lee Gardner. (New Hampshire Division of Forests and Lands.)

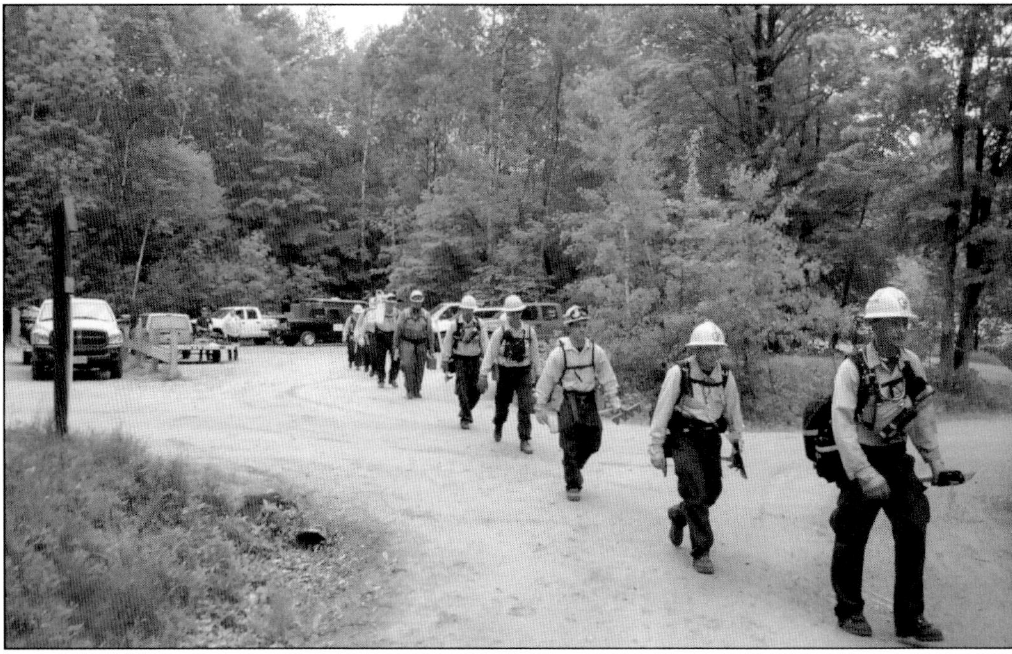

Members of this New Hampshire Wildland Fire Crew begin their shift on the Rattlesnake fire in the town of Rumney on May 29, 2008. The fire burned 54 acres of forest vegetation across sheer cliffs and ledges of the popular rock climbing area known as Rumney Rocks. Local, state, and federal firefighters spent three days extinguishing the wildfire. (New Hampshire Division of Forests and Lands.)

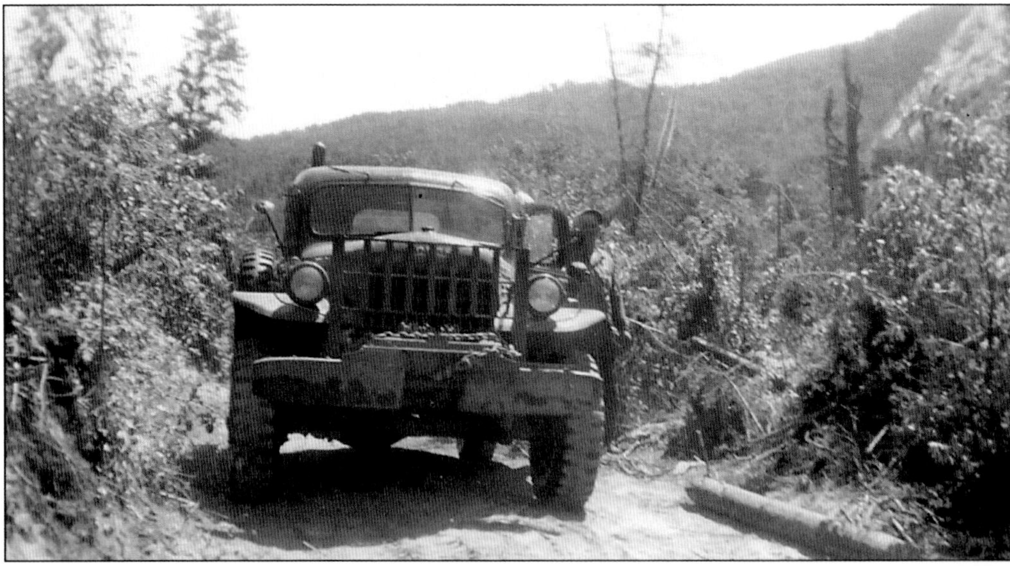

This 1952 photograph captures New Hampshire district chief Hartwell beside a Power Wagon at the Bald Cap wildfire. The fire burned on the steep slopes of the remote North Bald Cap Mountain in the unincorporated town of Success. Over the course of July and August, the fire charred 84 acres of spruce/fir forest; cost $37,162.62; and required firefighter assistance from the University of New Hampshire, Brown County, Franconia Paper Company, and soldiers from Camp Dodge. (New Hampshire Division of Forests and Lands.)

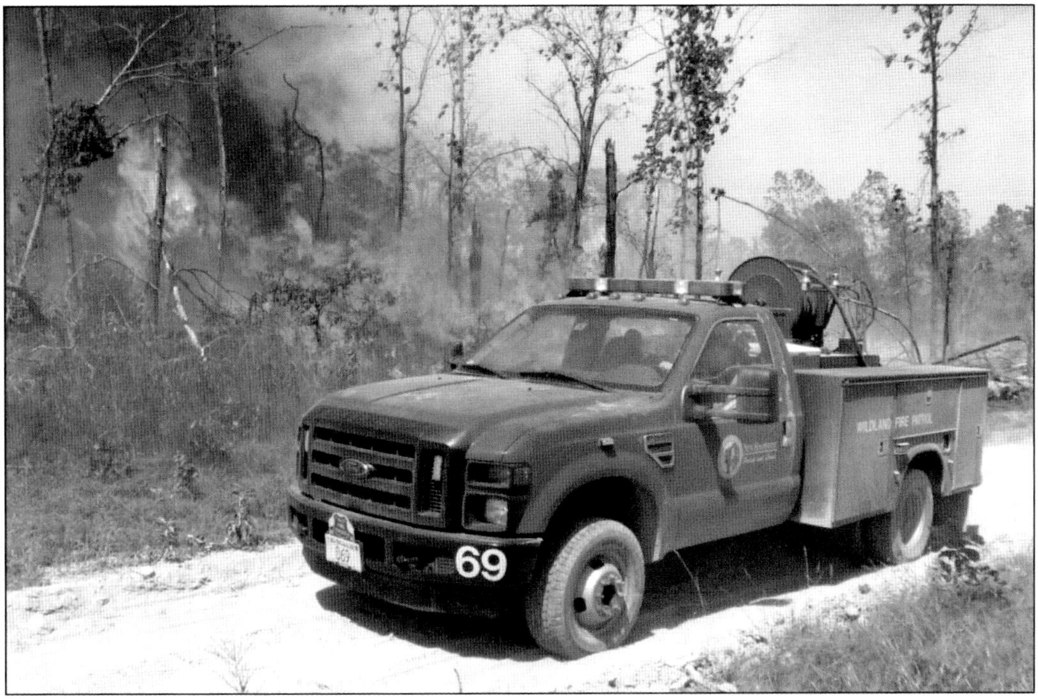

This New Hampshire Type 6 engine was the state's first engine to ever be deployed outside of New England in support of a national wildfire mobilization. During the early summer of 2012, the engine and crews were assigned for over 30 days to the Mark Twain National Forest in Missouri to support wildland firefighting operations on the forest. (New Hampshire Division of Forests and Lands.)

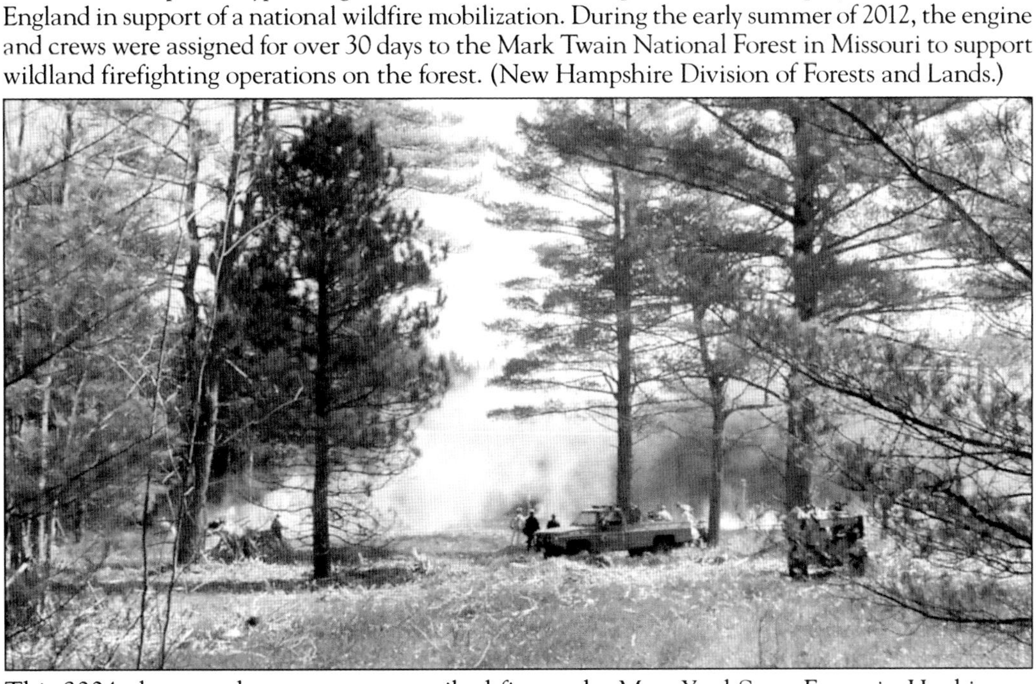

This 2004 photograph captures a prescribed fire at the Mast Yard State Forest in Hopkinton. New Hampshire has worked with federal, state, and local agencies to preserve wildlife habitats and unique ecosystems within the state through prescribed fires. The Mast Yard State Forest is named to historically recognize the area for its role in supplying ship masts to the shipbuilding industry of the 1700s and 1800s. (New Hampshire Division of Forests and Lands.)

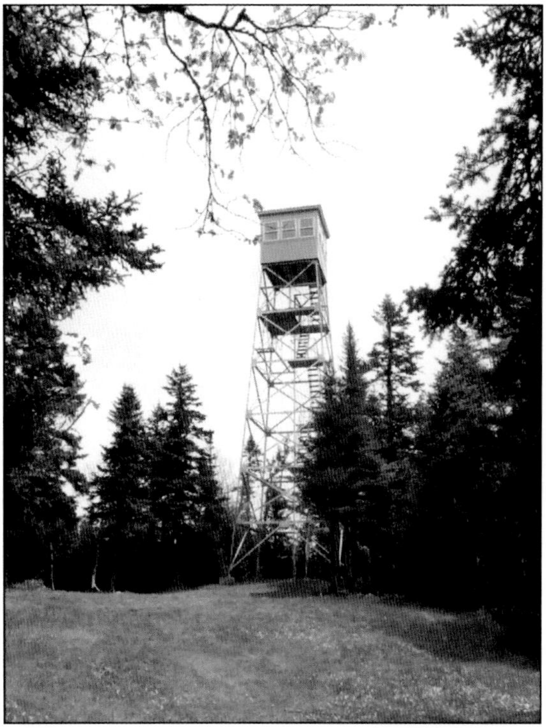

This is Magalloway Fire Lookout Station in 2009. This fire tower, located in the town of Pittsburg, is still in operation today. Starting in 2009, the New Hampshire Division of Forests and Lands has allowed citizens to rent out the watchman's cabin at the summit of the mountain, giving them a chance to briefly experience life as a watchman. Magalloway Fire Lookout Station is one of 15 stations in the state that is still staffed during periods of high fire danger. (New Hampshire Division of Forests and Lands.)

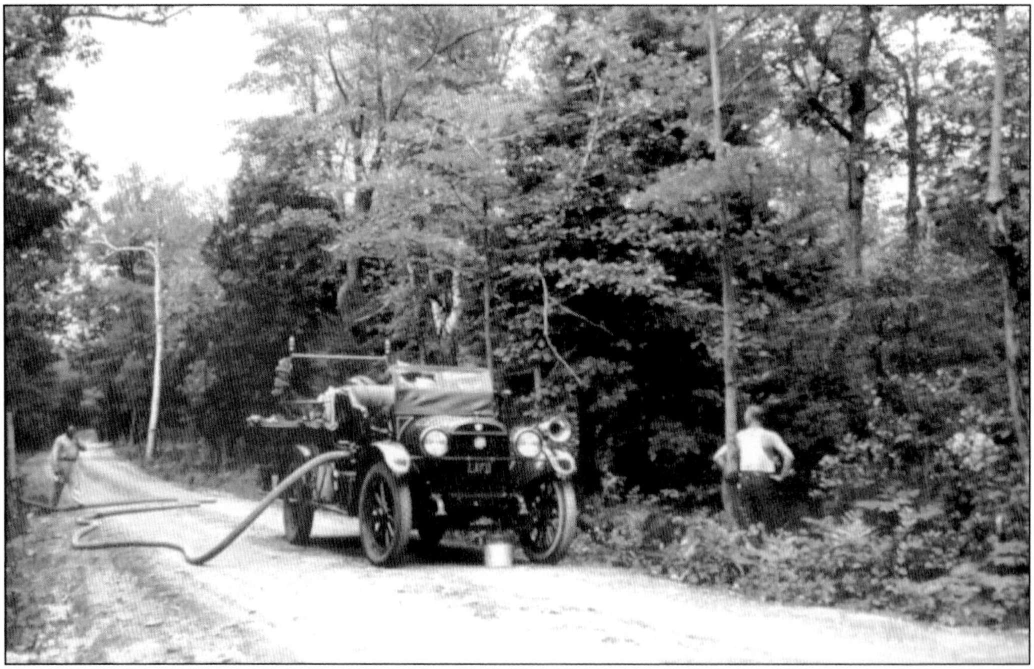

This 1930s photograph shows the Lost Acres Fire Department mopping up a wildfire in the town of Granby. The Connecticut General Statutes mandate that local fire departments are responsible for responding to and suppressing wildfires across the state. The Connecticut Division of Forestry supports wildfire suppression efforts as needed and provides wildfire suppression training to local fire departments. (Connecticut Division of Forestry.)

Connecticut Division of Forestry Fire Apparatus No. 22 was a late-1930s cab over a General Motors truck that was commercially constructed to serve its mission of fighting wildland fires. (Connecticut Division of Forestry.)

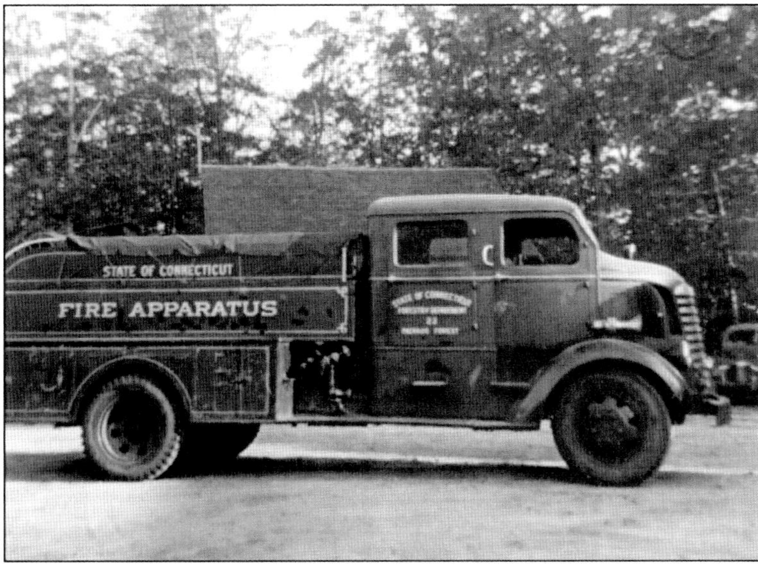

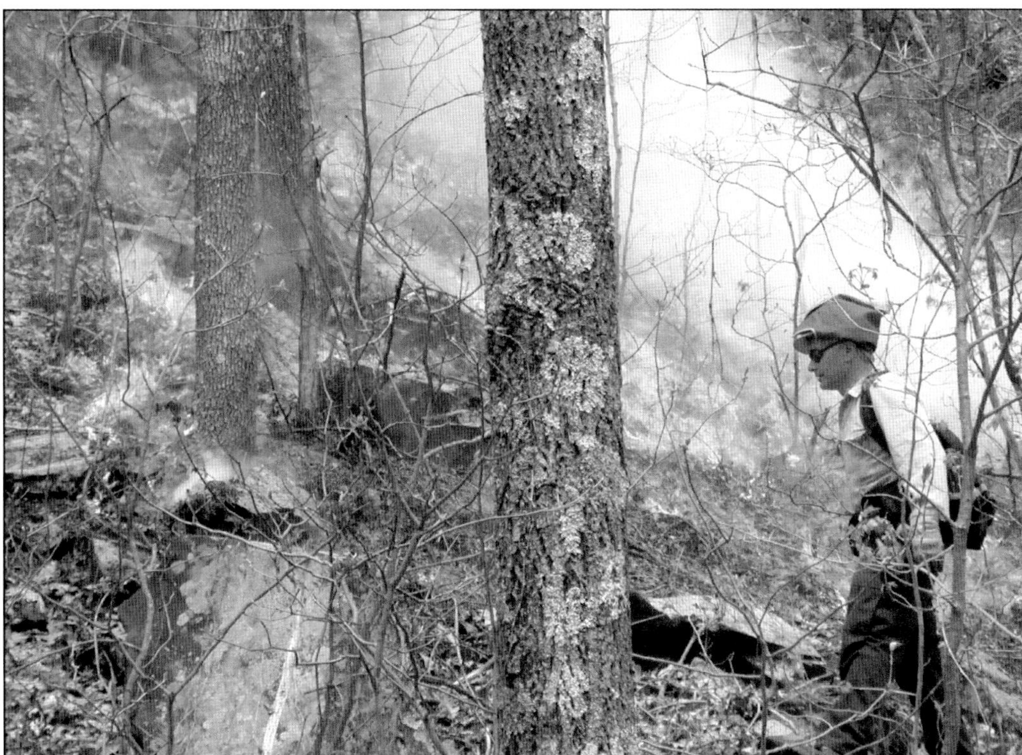

This 2003 photograph captures Connecticut forester Emery Gluck monitoring a prescribed fire on Soapstone Mountain at the Shenipsit State Forest in northeastern Connecticut. The Connecticut Division of Forestry maintains an active role in prescribed burning on state and private lands. In 1927, a forest fire lookout tower was established on Soapstone Mountain, and today, a public observation tower stands in its place. (Connecticut Division of Forestry.)

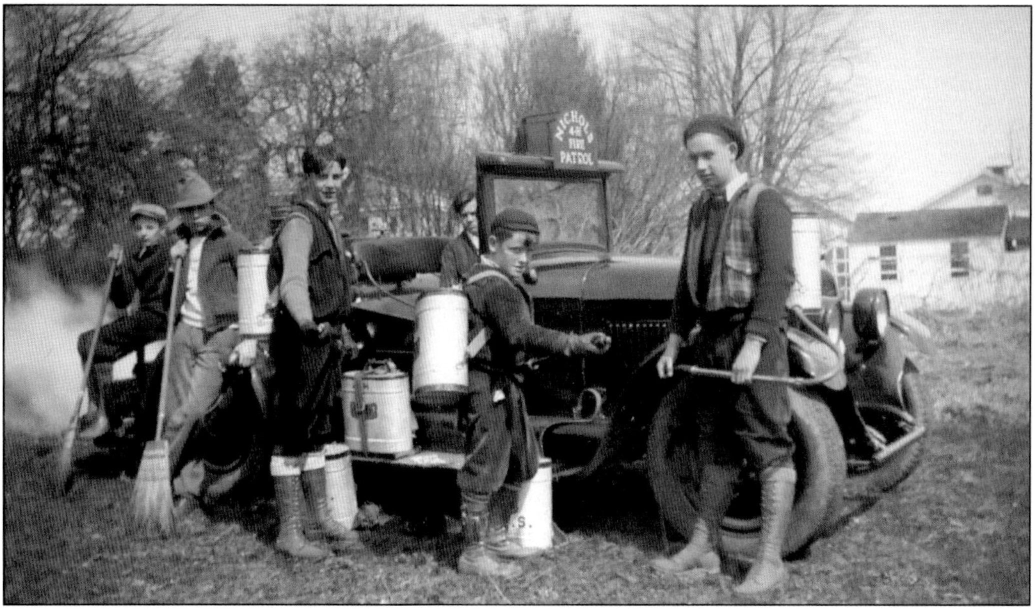

During World War II, there was a shortage of manpower to help fight forest fires across the nation, and high school students were recruited to help. This photograph captures the Nichols, Connecticut, 4-H Fire Patrol with its engine at the scene of a fire in Fairfield County. These types of crews were trained by Connecticut forest rangers, and they received two hours of training a week over a seven-week period. (Connecticut Division of Forestry.)

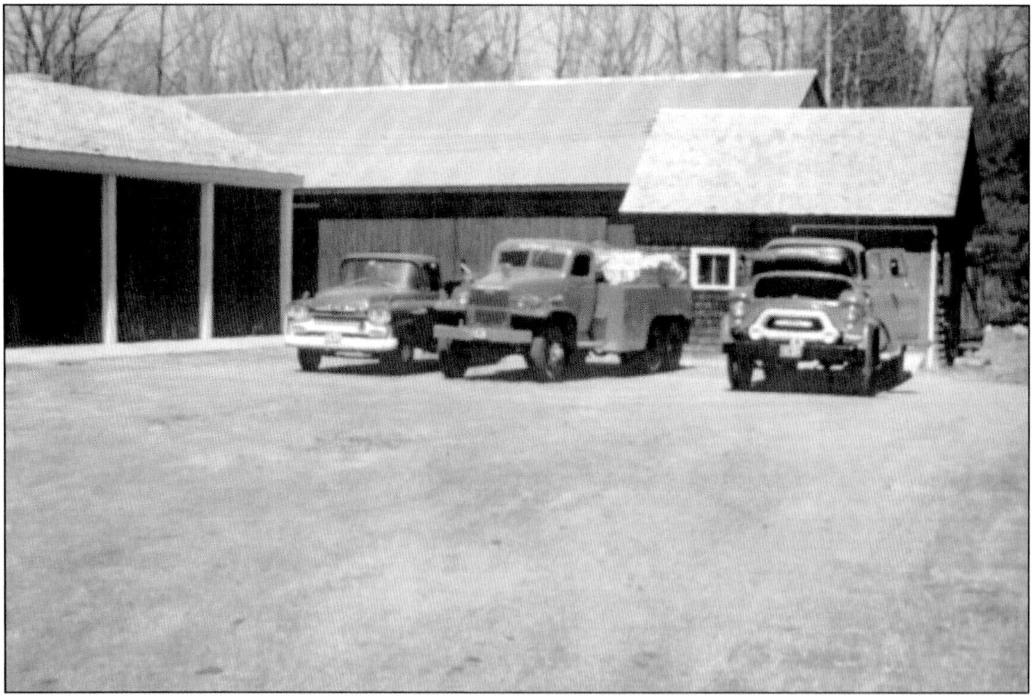

This photograph was taken during the spring of 1965 at the Pachaug State Forest headquarters near Voluntown. These three Connecticut Division of Forestry firefighting vehicles were stationed at the forest for a quick response to wildfires in the area. (Connecticut Division of Forestry.)

This Connecticut lookout tower was located on Mount Misery in the Pachaug State Forest. Once lookout towers played an important role in detecting wildfires across the state and dispatching firefighting resources. Today, all towers have been dismantled, and fires are reported through the state's 911 dispatch system. (Connecticut Division of Forestry.)

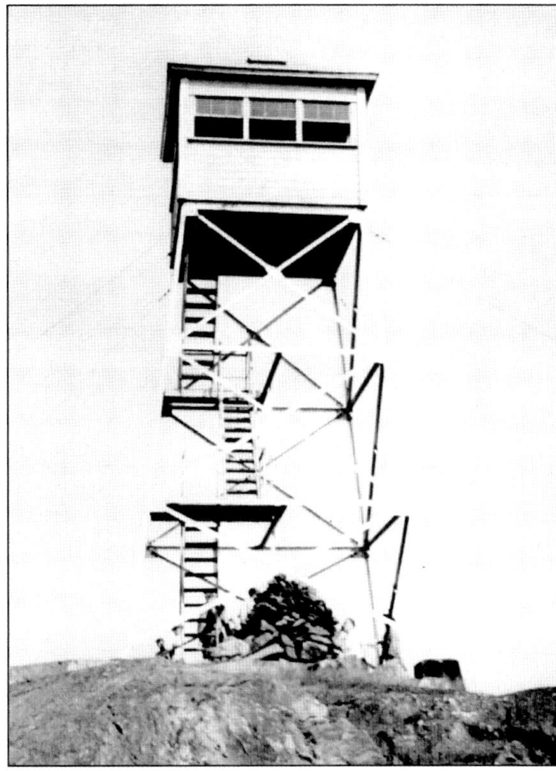

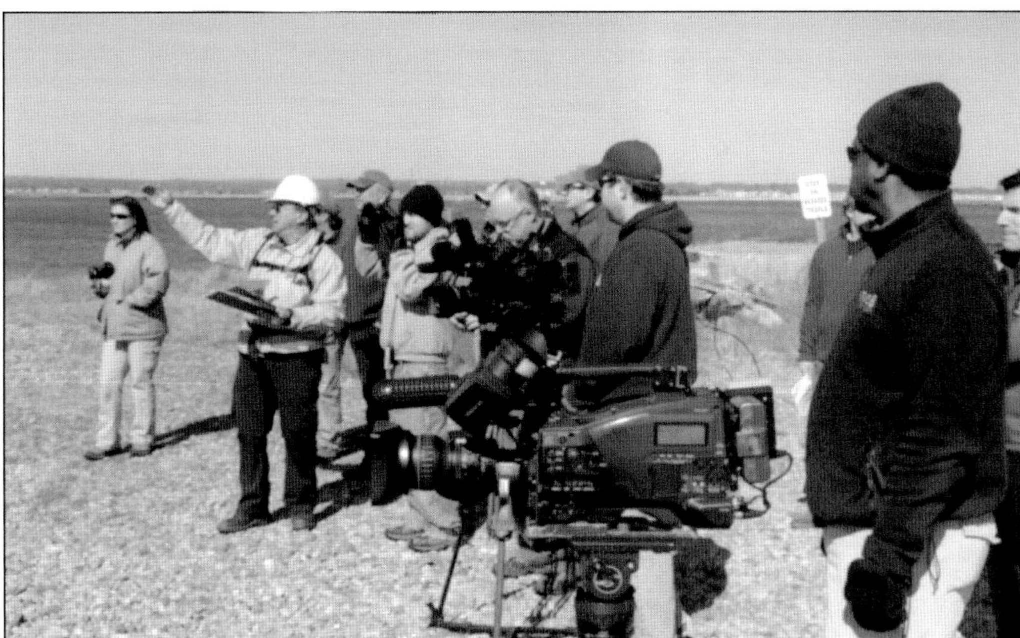

Connecticut State fire supervisor Ralph Scarpino, in wildland firefighting gear, provides an overview of an active prescribed burning project to news crews during its implementation by the Connecticut Division of Forestry in February 2012 along the Housatonic River. Scarpino served as Connecticut's state fire supervisor from 1990 through 2013. (Connecticut Division of Forestry.)

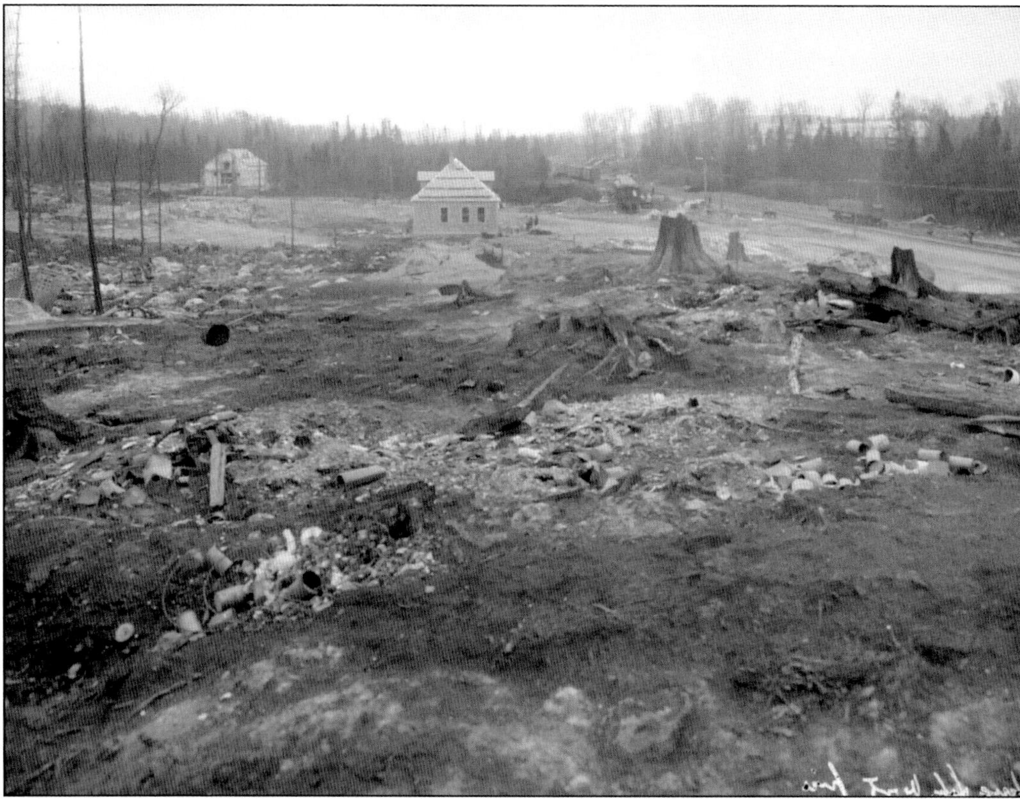

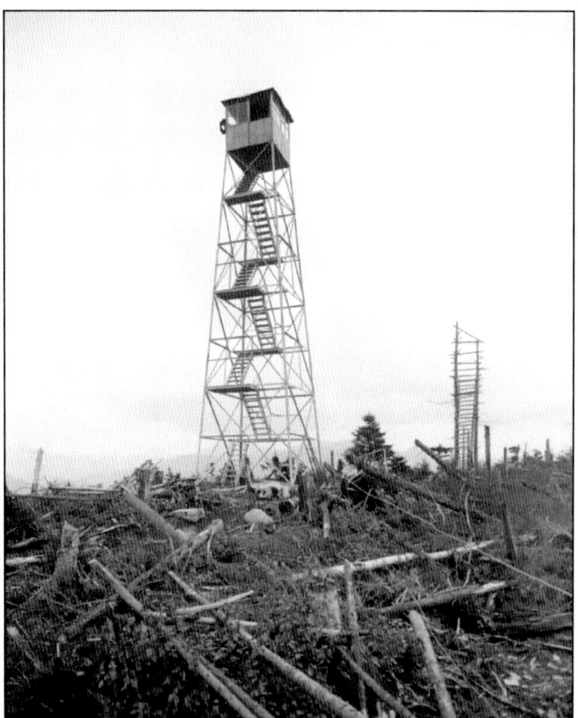

By 1885, the state of New York had suffered great damage from forest fires resulting from debris burning. Railroads, too, caused fires that burned vast areas of logged forests. The resulting severe floods and poor water quality led to the creation of the Adirondack and Catskill parks and a prohibition of timber harvesting on public lands. To protect these new state lands from wildfire, the state created a forest fire warden force to prevent and extinguish fires. (New York DEC State Forest Rangers.)

Beginning in 1905, New York's first fire towers were rustic wooden structures. Fire towers proved an effective means of detecting small forest fires, and from 1909 to 1964, a total of 115 steel towers were installed. By 1992, the state had decommissioned all state-operated fire towers, but many are still maintained by private groups. (New York DEC State Forest Rangers.)

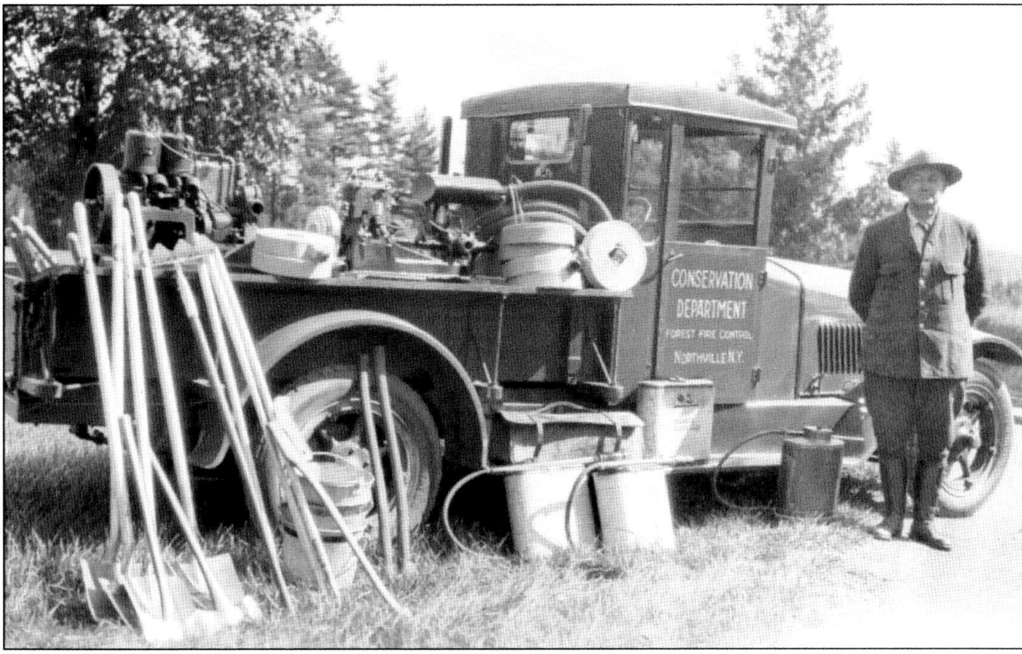

In 1912, the full-time professional position of forest ranger replaced the militia-type fire warden system. By 1929, vehicles had become the principle method of moving firefighting equipment where it was needed. Motorcycles were even used, some with built-in pumps so small fires could be contained quickly if a water source was nearby. Shovels, axes, pumps, and hoses are still the basic tools of forest firefighting. (New York DEC State Forest Rangers.)

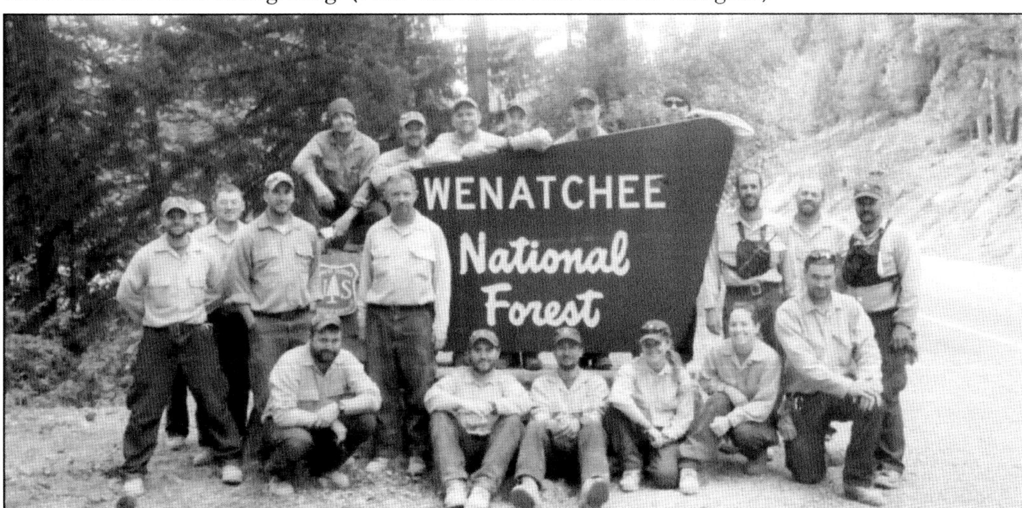

In 1970, the New York State Conservation Department became the Department of Environmental Conservation to include natural resource conservation and environmental protection. In 1979, New York sent its first two 20-person crews to western states to support national firefighting efforts. Since then, New York resources participate annually fighting fires in Florida, Texas, California, Oregon, Washington, and even Quebec, Canada. The training and experience of national firefighting contributes to New York's forest rangers' expertise in wildland fire. (New York DEC State Forest Rangers.)

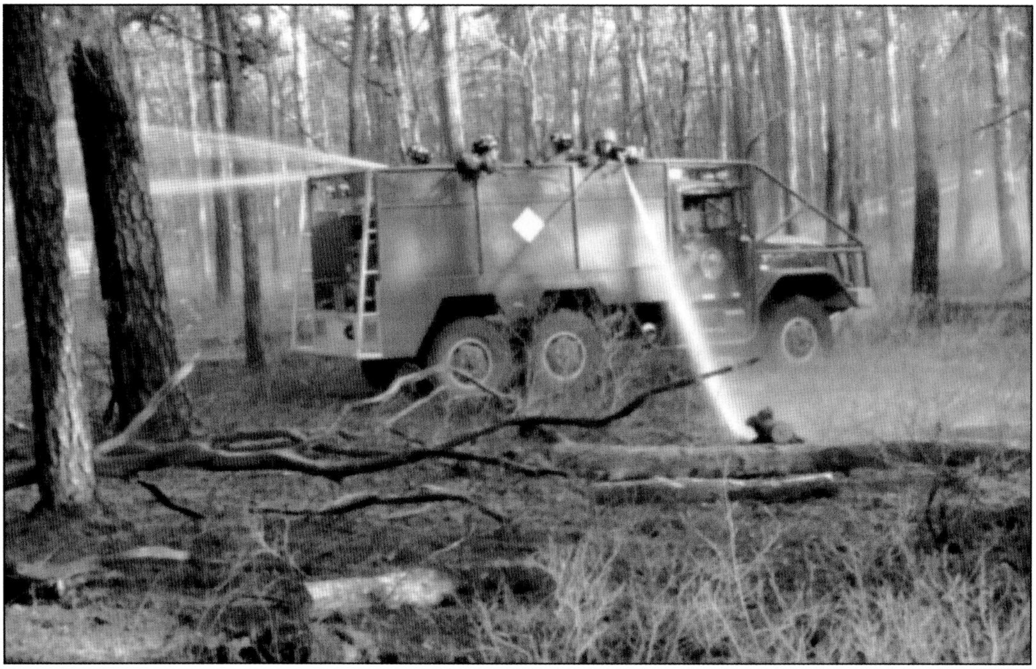

In 1995 and 2012, heavily suburbanized Long Island experienced dramatic wildfires that burned thousands of acres, destroyed numerous homes, and damaged much firefighting equipment. New York State forest rangers continue to support local firefighters, hosting a wildfire training academy every October. Today, several thousand local, state, and federal firefighters have been trained in wildland firefighting and incident command. (New York DEC State Forest Rangers.)

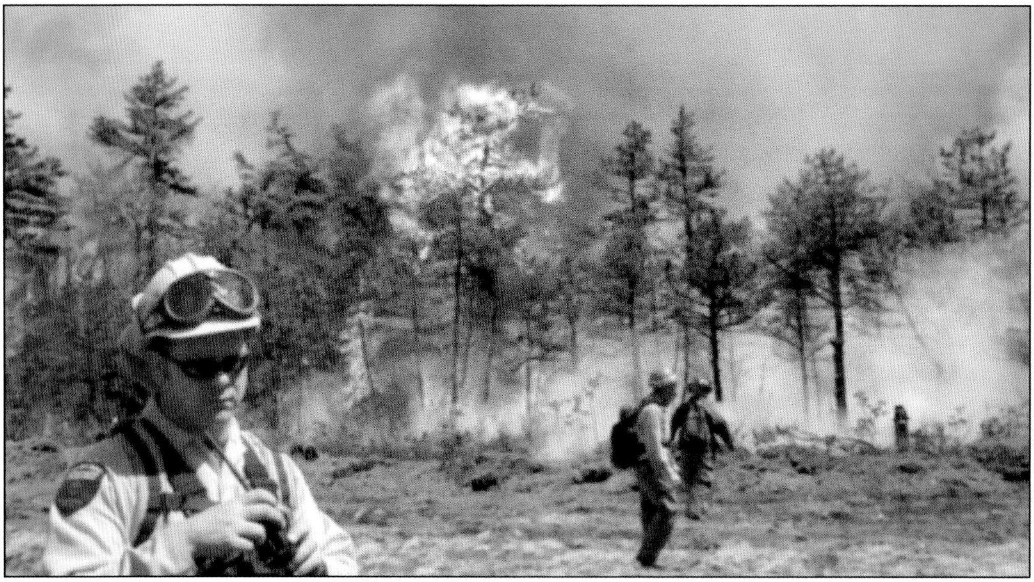

In 2008, a 2,800-acre wildfire burned through Minnewaska State Park in Ulster County, bringing to roadside view the destruction of public lands by wildfire. A 2010 change in open burning laws reduced spring fires by 37 percent, and for the first time in history, it caused debris burning to drop from the primary cause of wildfire to the second or third cause. (New York DEC State Forest Rangers.)

In 1964, New York forest rangers transitioned from a low band radio system to a VHF system with mountaintop repeaters. The state's mountainous terrain required a radio system that was not dependent on fire towers relaying messages. Along with the new radio system, aerial detection reduced the need to operate the system of 115 lookout towers. The 1960s were dry; only 95,142 acres burned as compared to 436,398 acres in the 1920s. (New York DEC State Forest Rangers.)

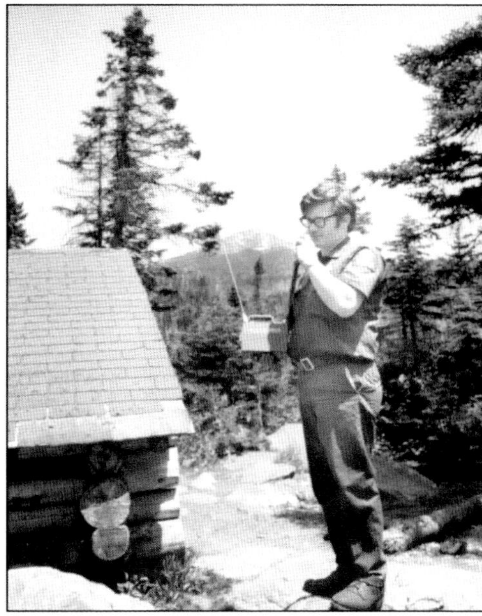

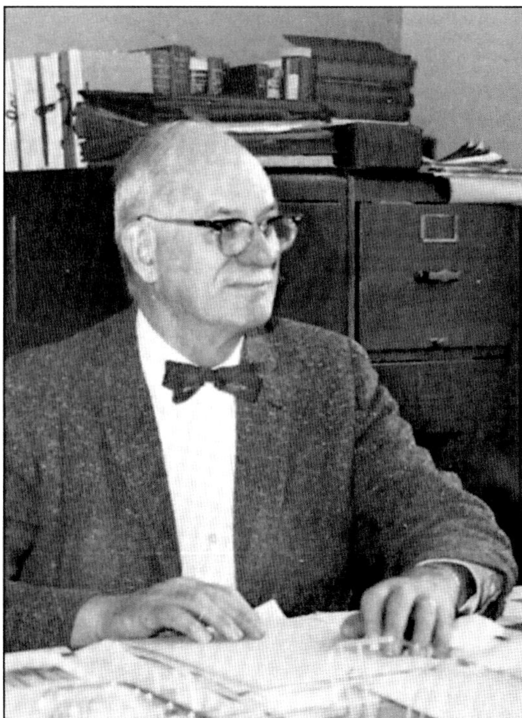

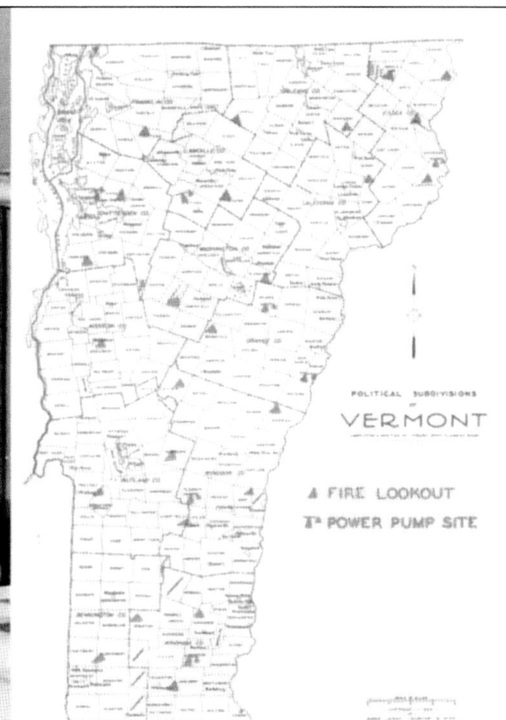

In November 1929, Perry Merrill was appointed as commissioner of forestry and began a long and influential career. Merrill served in forestry leadership roles until July 1966 and was responsible for making the Vermont Forest Service a prominent and respected state agency. Through his efforts with the CCC, this program constructed seven steel and three wooden lookout towers and suitable cabins for lookout watchmen and installed better telephone service. (Vermont Department of Forests, Parks, and Recreation.)

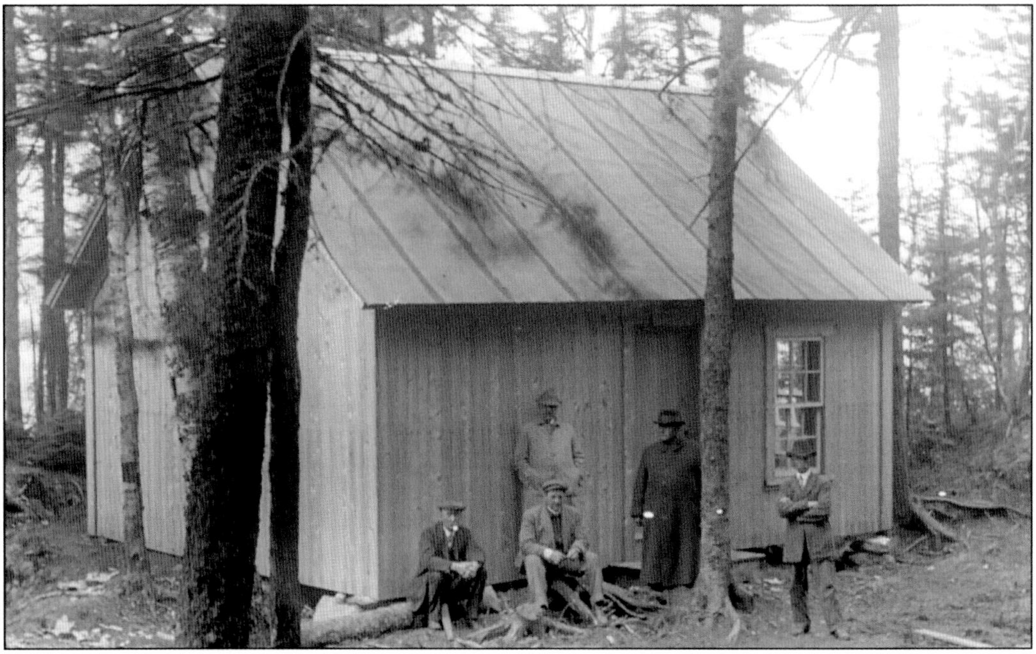

The first two fire lookout towers were constructed on Burke Mountain and Gore Mountain in northeastern Vermont. The lookout towers were in service for the 1913 fire season and were established by private woodland owners. These two towers, plus several other lookout stations, and forest fire patrolmen comprised the early fire detection system in Vermont. This photograph shows the lookout's cabin on Burke Mountain just after completion in 1912. (Vermont Department of Forests, Parks, and Recreation.)

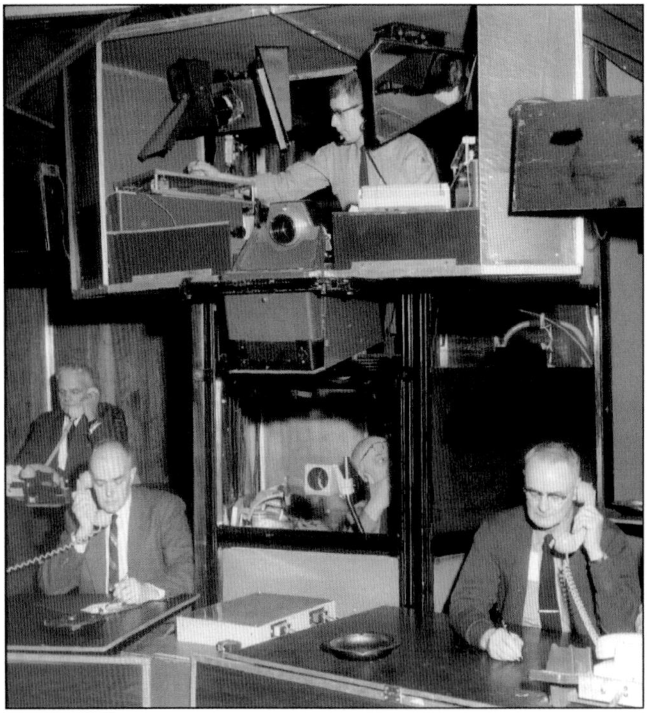

Vermont is one of the original seven members in the Northeastern Forest Fire Protection Commission (NFFPC), the first of its kind in the United States. This fire protection compact was initiated following the devastating forest fires throughout New England in the fall of 1947 and was signed into law by Vermont in 1949. The State of Vermont continues to play an active role in the NFFPC. Pictured is the 1966 NFFPC training session for "Simulated Fire Control Involving Fire Behavior Techniques." (Vermont Department of Forests, Parks, and Recreation.)

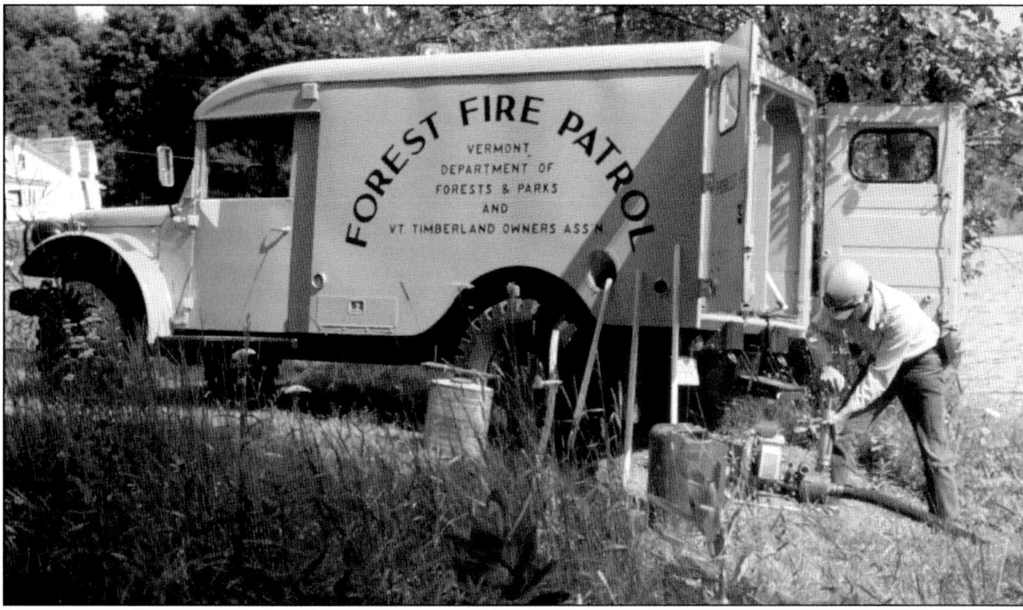

Through the 1970s, the Vermont Department of Forests, Parks, and Recreation acquired military vehicles and equipment through the Federal Excess Personal Property program. This equipment was loaned to towns through cooperative agreements and has greatly assisted small rural fire departments in forest fire suppression. The State of Vermont also used Federal Excess Property for Patrol Trucks and water tenders from the 1970s through the early 2000s. (Vermont Department of Forests, Parks, and Recreation.)

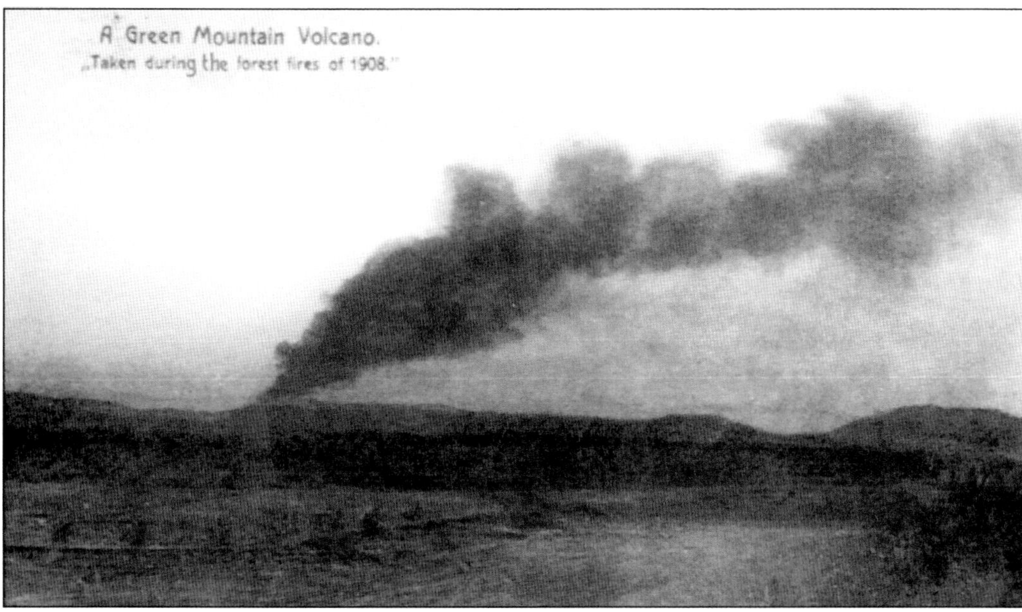

In December 1904, the Vermont Legislature created the Town Forest Fire Warden system and a forestry commission after large forest fires burned throughout the state in 1903. These laws laid the foundation for creation of the Vermont Forest Service in 1908 after an estimated 20,000–25,000 acres burned in the state that year. This photograph shows a large forest fire burning in North Troy, Vermont, in 1908. (Vermont Department of Forests, Parks, and Recreation..)

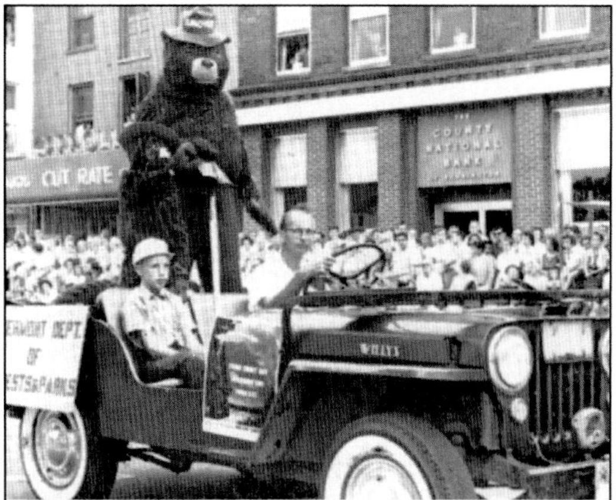

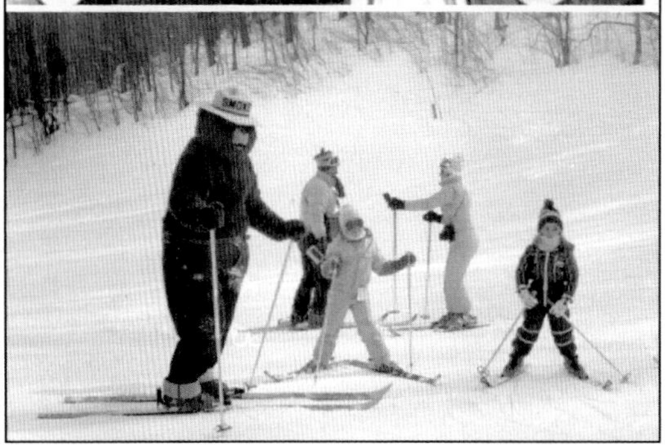

Smokey Bear has been part of Vermont's fire prevention message for decades. Smokey Bear and Lil Smokey appear in a parade in Bennington around 1960, and Smokey hits the slopes in the mid-1980s, enjoying Vermont's favorite winter pastime. Smokey Bear continues to make appearances throughout Vermont at fire-related events, state fairs, and school functions. A strong fire prevention message and the forest fire warden system have successfully kept wildland fires to a minimum in Vermont. (Vermont Department of Forests, Parks, and Recreation.)

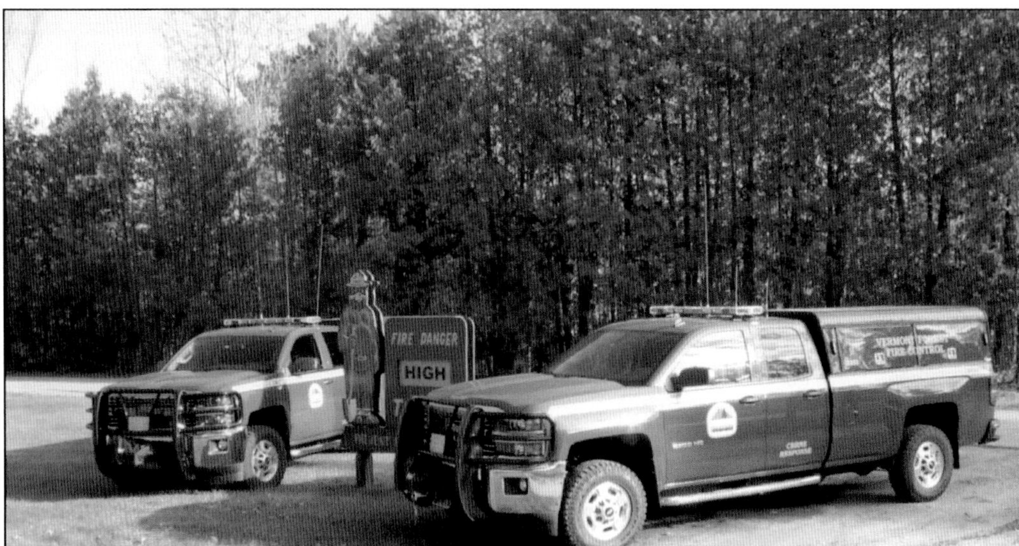

Today, the Vermont Department of Forests, Parks, and Recreation has two dedicated forest fire response vehicles and four fire trailers and maintains an inventory of wildland fire pumps, hand tools, and radios. Vermont's wildland fire program focuses on training and working with regional, federal and state partners in fire prevention, education, suppression, and predictive services. The Town Forest Fire Warden system continues to be the backbone of Vermont's forest fire program at the local level with over 275 appointed fire wardens. (Vermont Department of Forests, Parks, and Recreation..)

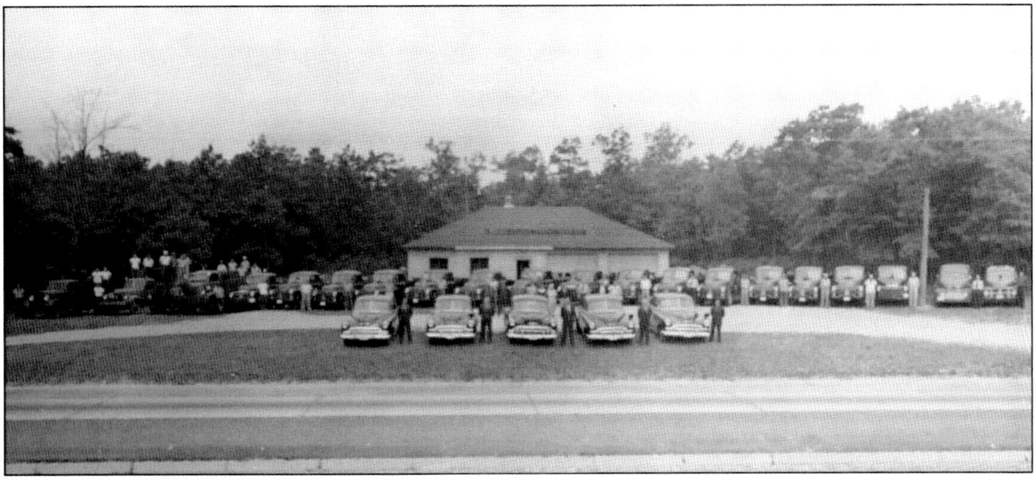

This 1953 photograph captures the firefighting fleet of the Rhode Island Forest Service in front of the headquarters building at Dawley Memorial State Park. The Rhode Island Commission of Forestry was founded in 1906, and the first forest wardens were appointed in 1909 under the direction of Commissioner Dr. Jesse B. Mowry. (Rhode Island Division of Forest Environment.)

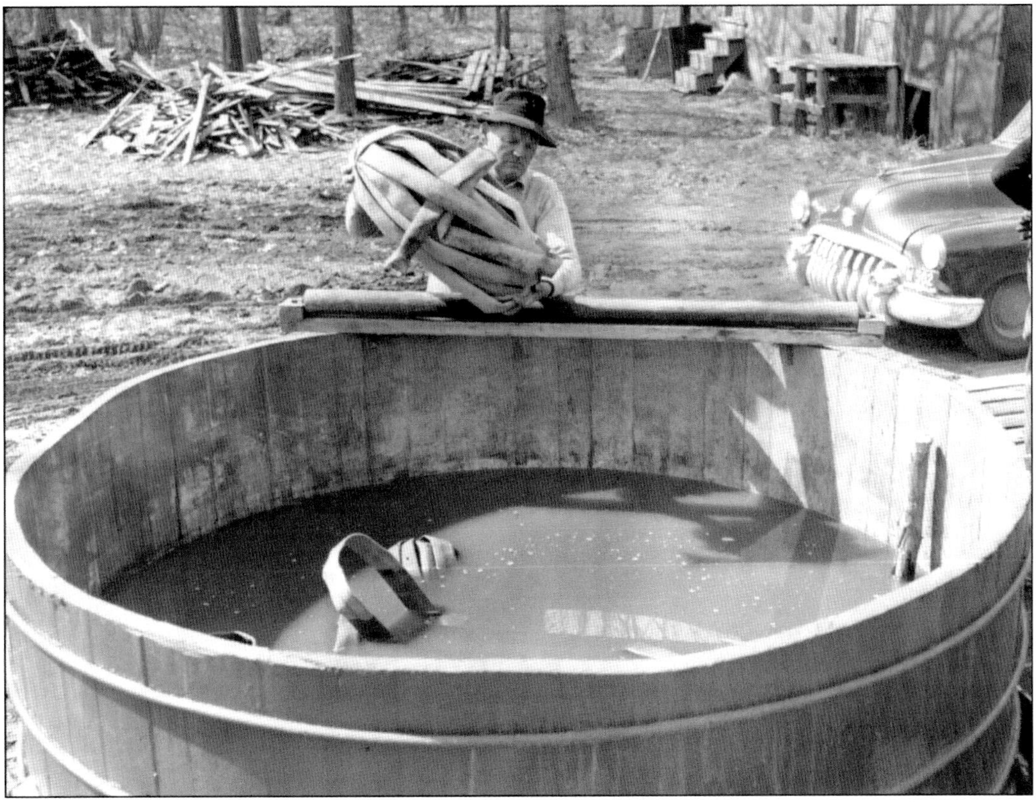

Rhode Island Forest Service foreman Edmund Rush dumps hose into the first hose washing station developed in 1956. At this time, Rhode Island initiated its equipment on loan program, which would provide hose to the state's local fire departments for use in suppressing wildfires. The initial authorization of the program provided funds to the Forest Fire Equipment Committee for the purchase of 20,000 feet of linen hose. (Rhode Island Division of Forest Environment.)

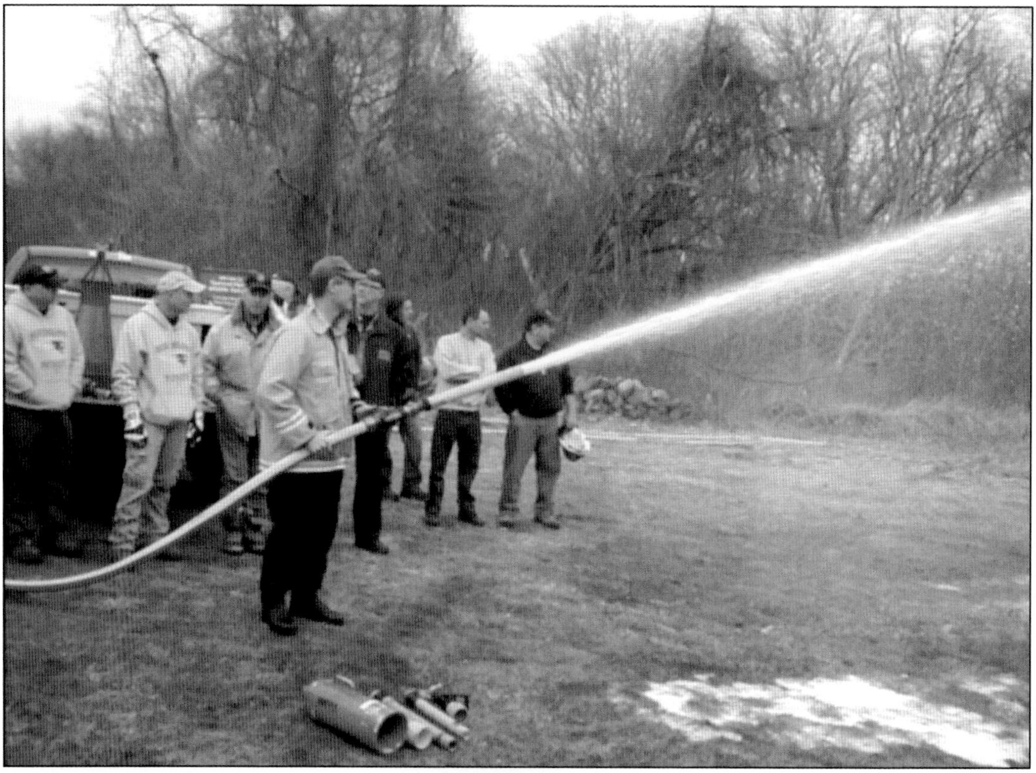

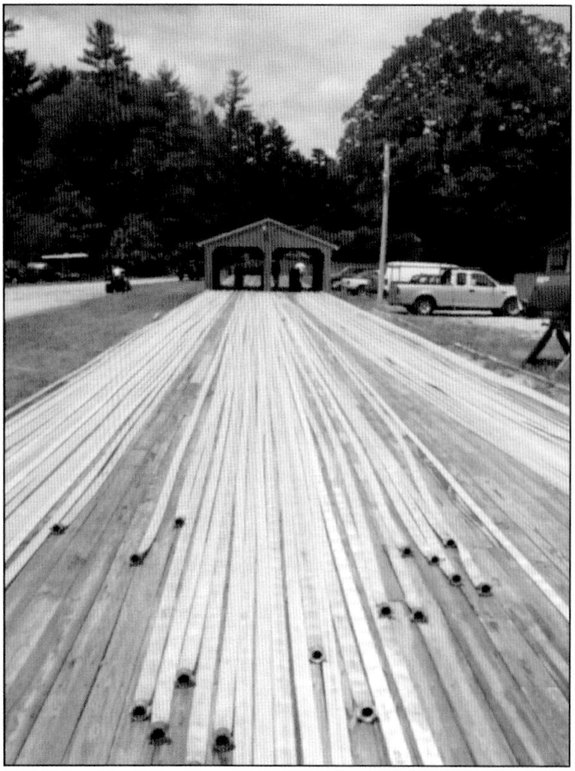

Training local firefighters has been one of the priorities of the Rhode Island Division of Forest Environment. This photograph taken at a Class A foam training class shows Capt. John Dodge, of the New Hampshire forest rangers, instructing a hands-on demonstration at the Warren Fire Department. (Rhode Island Division of Forest Environment.)

This photograph shows the hose processing deck at the Rhode Island Division of Forest Environment's Arcadia headquarters in 2015. Today, 35 miles of 1.5-inch weeping hose with quarter-turn connections is on loan to Rhode Island's 67 local fire departments through their equipment on loan program. (Rhode Island Division of Forest Environment.)

Today, the Rhode Island Division of Forest Environment leads the way in forest protection and conservation efforts across the state. It maintains a firefighting fleet that includes a strike team of wildland engines that is available for response to wildfires, as well as being utilized to support prescribed fire, risk reduction, and prevention activities. (Rhode Island Division of Forest Environment.)

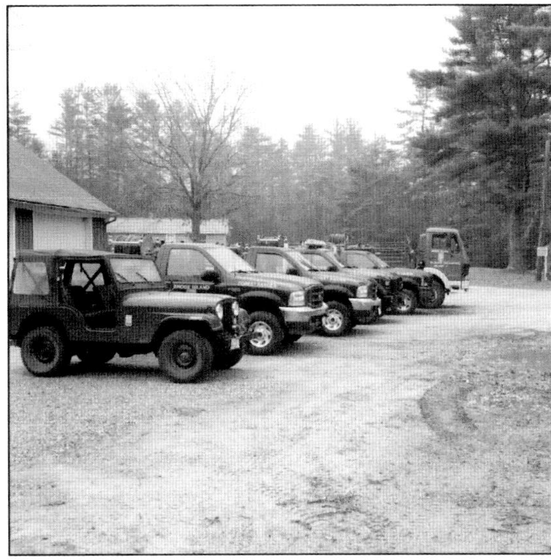

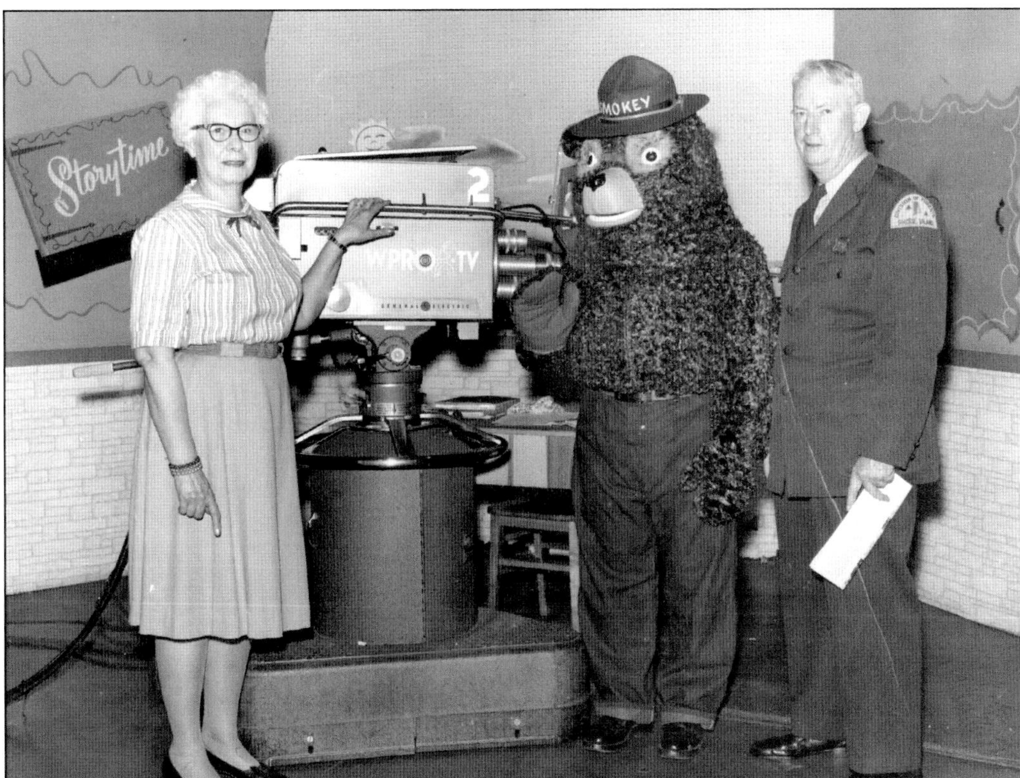

While prevention outreach has always been a cornerstone of the Rhode Island Division of Forest Environment, the fire prevention work accomplished by Ranger George E. Matteson is of considerable contribution. In 1958, Ranger Matteson conducted 150 forest fire prevention talks that reached over 11,000 children and an additional 5,000 adults. He also made 12 television appearances with Smokey Bear and had numerous radio and newspaper interviews to his credit. (Rhode Island Division of Forest Environment.)

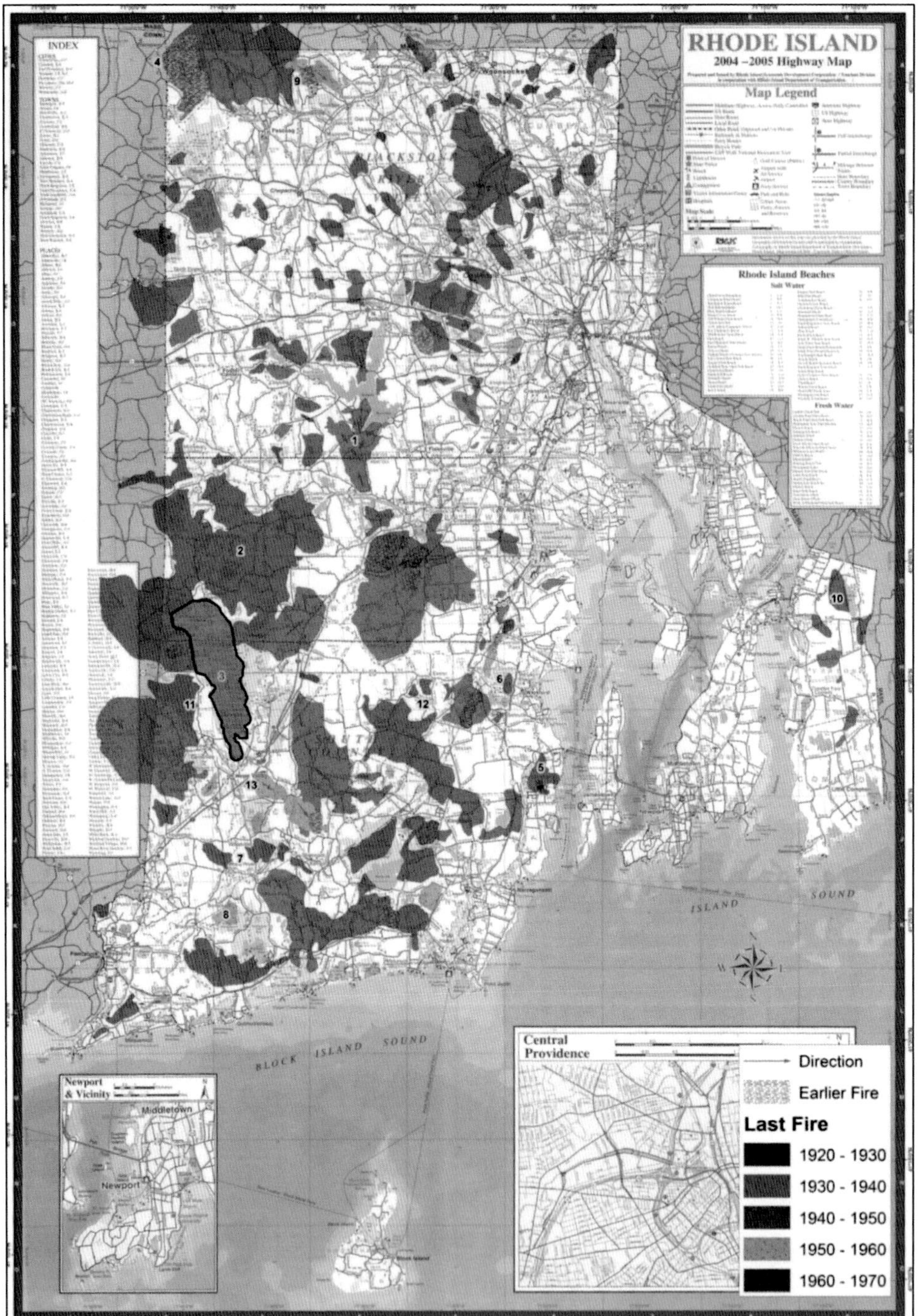

This map was created by Ranger George Matteson and shows the major fire history of Rhode Island from 1910 through 1960. One of the most notable fires was the Coventry fire, which burned 24,510 acres, destroyed 14 homes and 6 camps, and injured 180 individuals during May 1942. In May 1930, a series of wildfires burned 34,700 acres. The Wood River Valley fire blew up on May 2, 1951, burning 7,300 acres and resulting in the death of one firefighter and critically injuring seven others because of firefighter entrapments. (Rhode Island Division of Forest Environment.)

Three

Mid-Atlantic States

The Mid-Atlantic Interstate Forest Fire Protection Compact (MAIFFPC) is comprised of Delaware, Maryland, New Jersey, Pennsylvania, West Virginia, Ohio, and Virginia. Virginia is not within the northeastern states but is a member of the MAIFFPC. In March 1952, an interstate conference on cooperative fire protection was held at Harrisburg, Pennsylvania, under the auspices of creating a forest fire compact for the mid-Atlantic states. States passed enabling legislation to enter a compact between 1953 and 1967. The US Congress approved the creation of the Mid-Atlantic Interstate Forest Fire Protection Compact in 1956. The first official meeting was held in 1965 where the first bylaws were developed. In 1969, the US Forest Service officially assigned W.M. Stiteler as a liaison. He served as the first compact coordinator. Ohio joined in 1988, becoming the seventh member.

One of the main responsibilities of the compact is to provide for the maintenance of a regional forest fire plan for the member states. Since its creation, the compact has provided a medium to exchange ideas and experiences. It has fostered the idea of "get together," as described by Maryland State forester A.R. Bond at the 1970 training meeting of the compact. Over the years, one of the highest priorities of the compact has been to serve as a focal point for training firefighters and fire managers among the member states. The first training meeting of the compact was held in 1966 at Indiantown Gap, Pennsylvania. Many training meetings were hosted at the National Fire Academy in Emmitsburg, Maryland, due to its central location for easy travel by all member states. In 2007, the MAIFFPC created an annual Wildfire Academy in partnership with West Virginia University and has delivered basic and advanced wildland firefighting training courses to individuals from across the region and beyond.

The states within the MAIFFPC are varied in their topography and fuels, as well as the unique fire protection and management organizations that they maintain within their states. The Mid-Atlantic states stretch from sandy plains along the Atlantic Ocean with their fire-adapted vegetation, such as pitch pine, to the hilly and rugged terrain of the interior with hardwood forests. All states exhibit wildland urban interface areas that challenge the state fire managers in providing for the protection of life, property, and natural resources from the threat of wildfire.

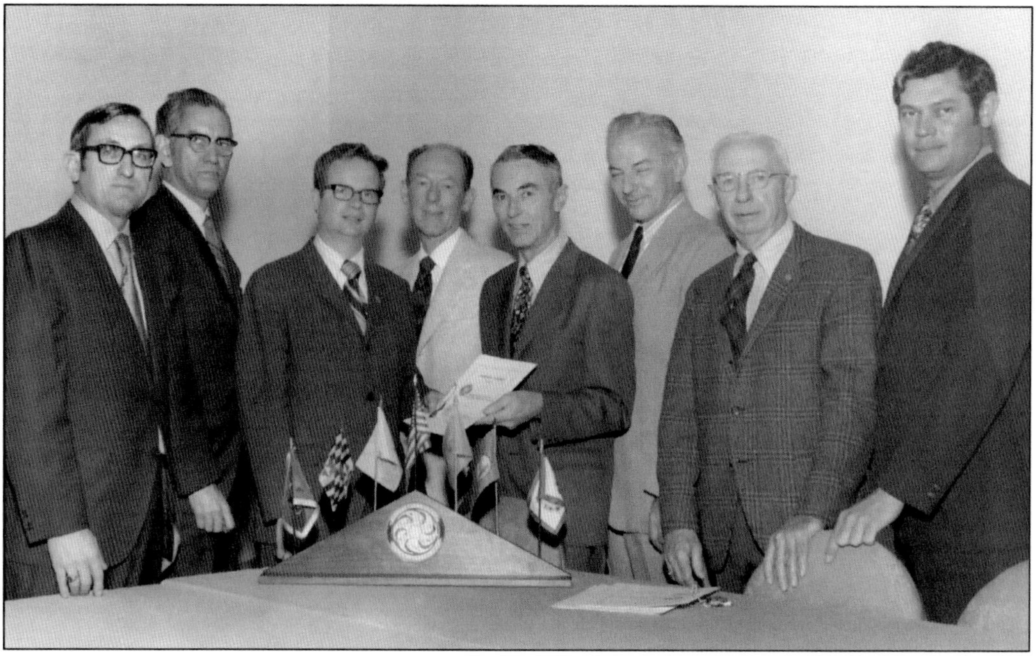

On February 2, 1972, an agreement expanding mutual aid assistance in wildland firefighting and training was signed between the MAIFFPC and Southeastern States Forest Fire Protection Compact. This was the first inter-compact agreement ever signed. Attending the signing were, from left to right, Philip Archibald (USFS), and state foresters Samuel Cobb (Pennsylvania), Samuel Mace (Delaware), George Moorhead (New Jersey), Lester McClung (West Virginia), A.R. Bond (Maryland), George Dean (Virginia), and Ronald Schureman (Kentucky). (MAIFFPC.)

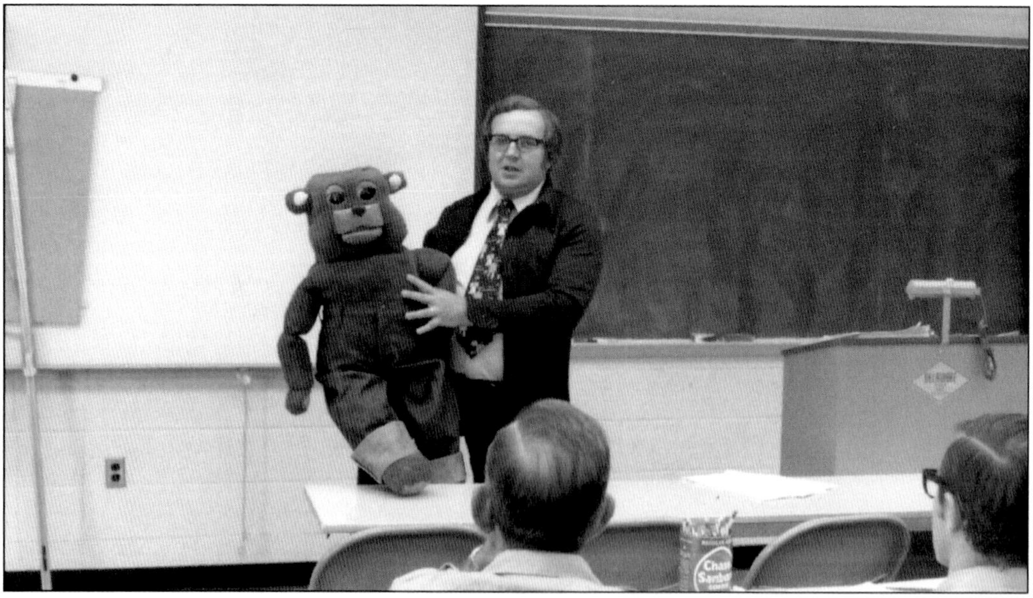

The 1976 MAIFFPC meeting was hosted by the Delaware Forest Service at the Delaware State Fire School. The theme of the meeting was the analysis of large and problematic wildfires. In this photograph, Art Kreelman, from the Pennsylvania Bureau of Forestry, discusses wildfire prevention activities along with a hatless Smokey Bear doll. (MAIFFPC.)

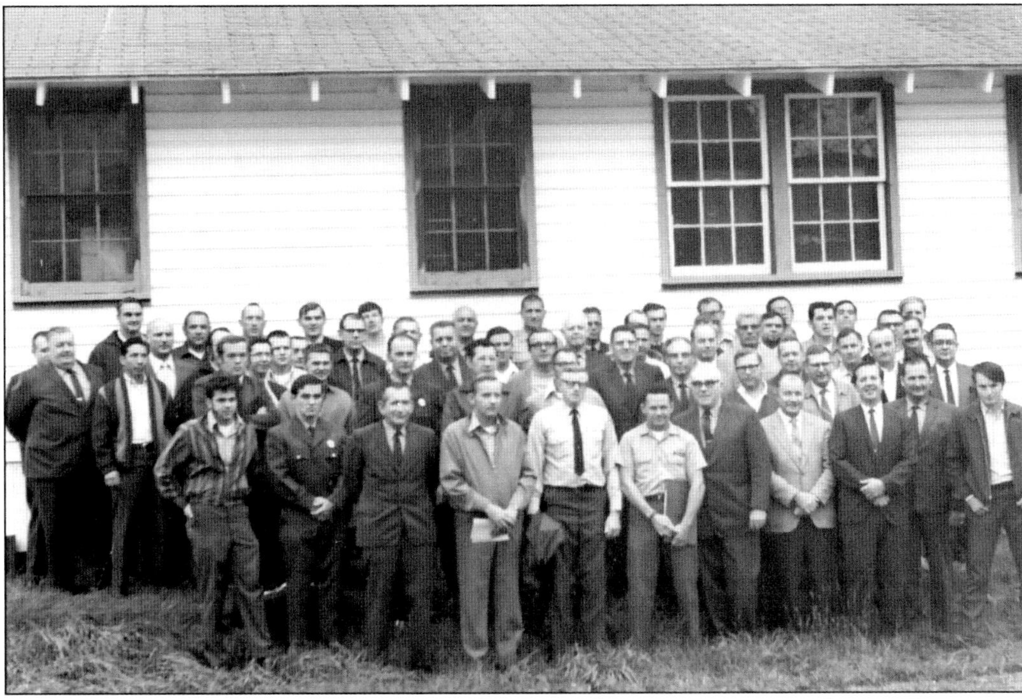

The sixth meeting of the MAIFFPC was held in Fort Indiantown Gap, Pennsylvania. The meeting was hosted by the state of Delaware, and large fire organization was the training topic. This photograph captures the attendees from the member states and was taken in front of one of the military barracks. The first training meeting was hosted by Pennsylvania in 1966. (MAIFFPC.)

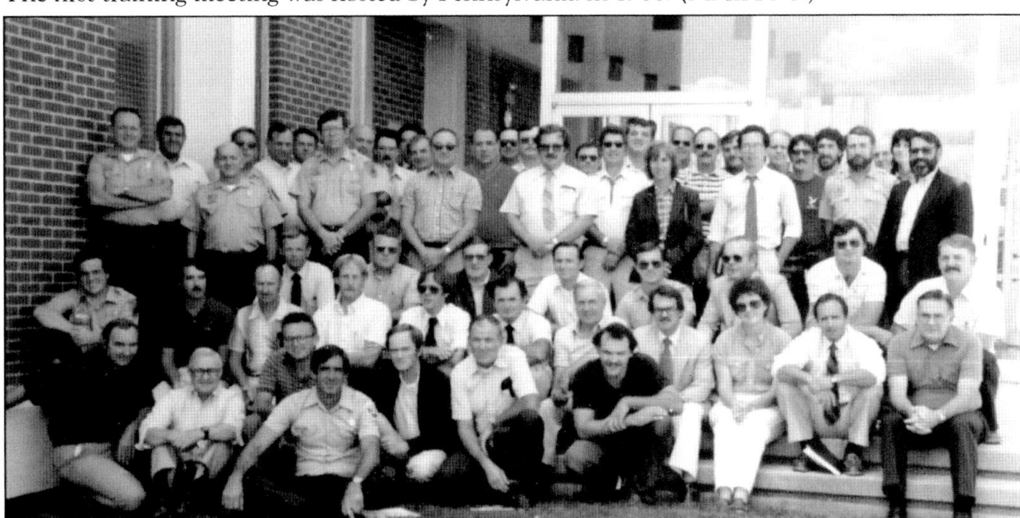

In 1980, the Mid-Atlantic Interstate Forest Fire Protection Compact partnered with the National Fire Academy (NFA) in Emmitsburg, Maryland. The NFA agreed to host the compact's annual training session. Over a dozen annual training meetings were held at the fire academy. This photograph captures attendees at one of the early-1980s meetings outside of the facility. The invite to the 1980 meeting requested that all participants wear coats and ties during the formal sessions. (MAIFFPC.)

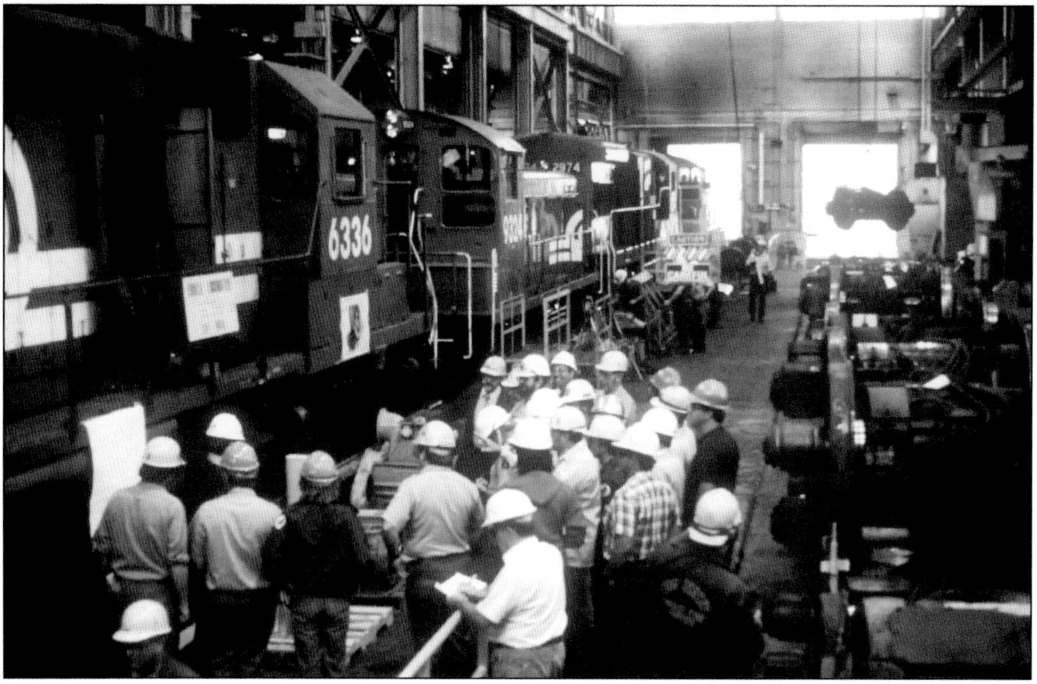

The 1985 meeting of the compact was hosted by the state of Delaware, and railroad fire prevention was the meeting's theme. In this photograph, compact members learn about diesel locomotives and their potential for igniting wildfires at a Conrail railroad train maintenance facility. (MAIFFPC.)

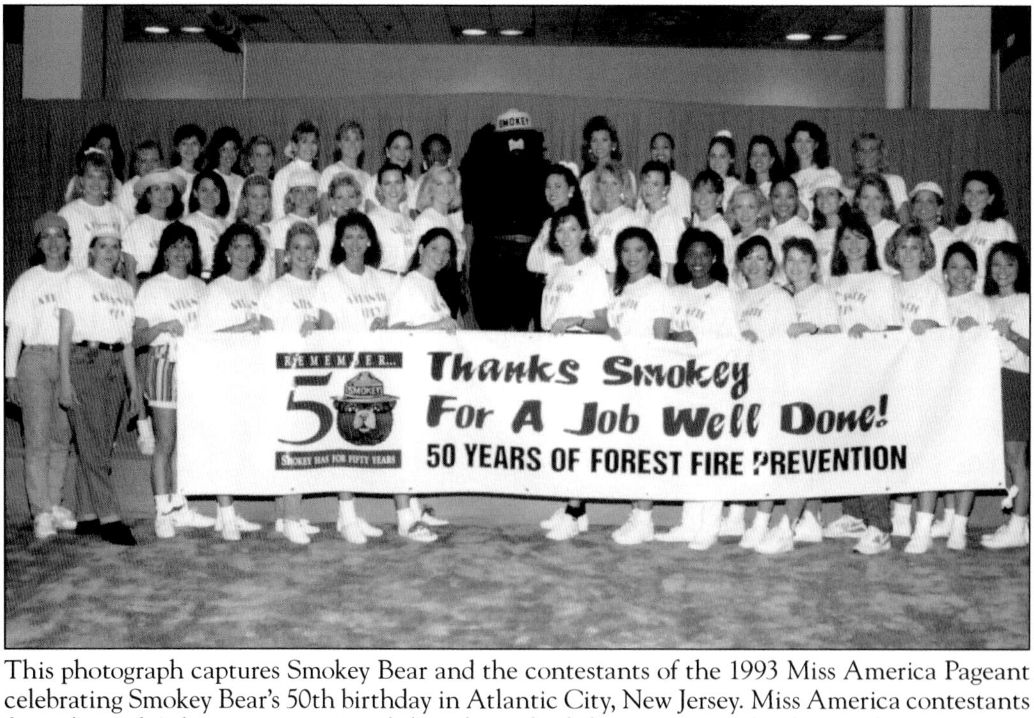

This photograph captures Smokey Bear and the contestants of the 1993 Miss America Pageant celebrating Smokey Bear's 50th birthday in Atlantic City, New Jersey. Miss America contestants from the mid-Atlantic states assisted their forest firefighting agencies by sharing fire prevention messages as they met with schoolchildren throughout the year. (MAIFFPC.)

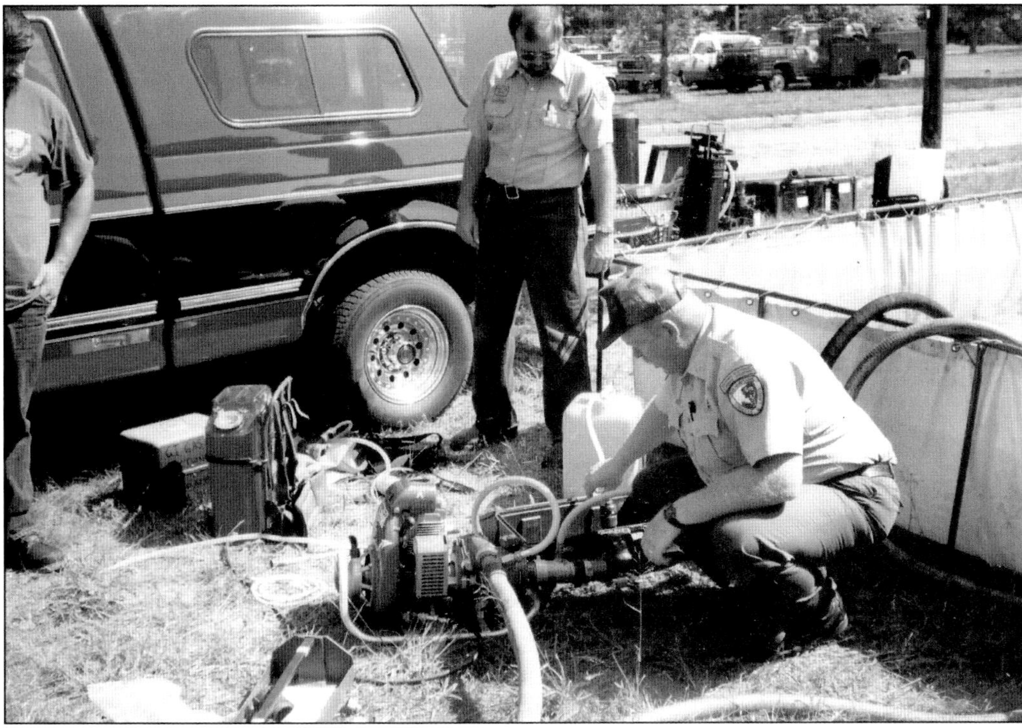

During the 1990s, a priority of the compact was to provide Class A foam training to all member states. A cadre of instructors was developed that could travel among the states and provide basic Class A foam use training with hands-on live fire evolutions. In this photograph, Jim Petrini (standing), of New Jersey, and Sven Carlson (kneeling), of New Hampshire, check over their pump and foam equipment. (New Jersey Forest Fire Service.)

The New Jersey Forest Fire Service was established by law on April 18, 1906, (Chapter 123, Laws of 1906), and the act became effective on July 4, 1906. This photograph captures Larry Terhune (an early New Jersey wildland firefighter) on fire patrol in 1914. Terhune was 20 years old and was compensated $60 a month for himself and his horse, plus 10¢ per hour for fighting forest fires, with funds provided by the federal Weeks Act. (New Jersey Forest Fire Service.)

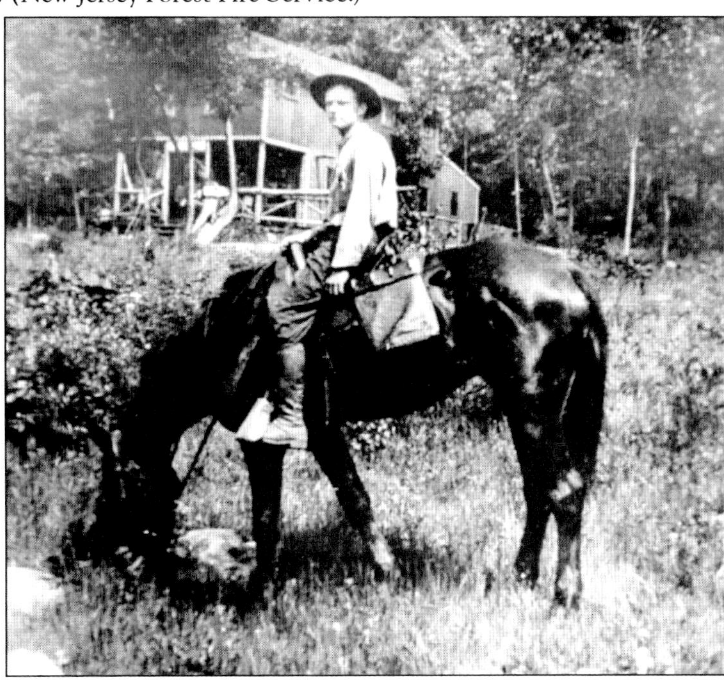

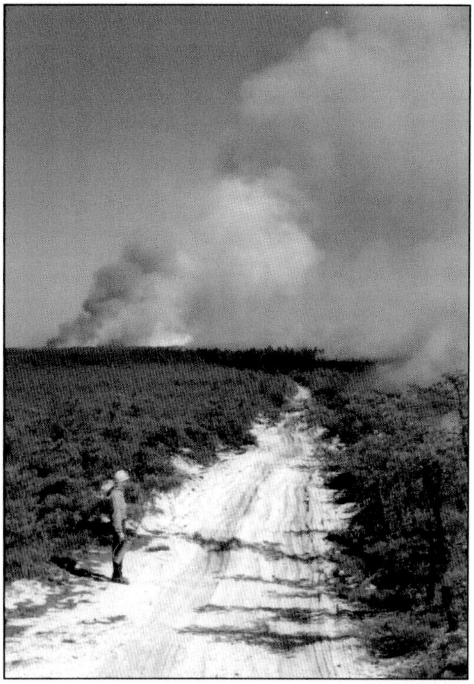

The pine plains forests that occur in the New Jersey Pine Barrens are considered globally significant, and the state's dwarf pitch pine plant communities are the largest in the world. The two largest sites include the East Plains and the West Plains. Most scientists believe that the long history of frequent and severe wildfires has been responsible for the creation and management of this unique forest vegetation. Wildfires that occur in this fuel type are explosive in growth. (New Jersey Forest Fire Service.)

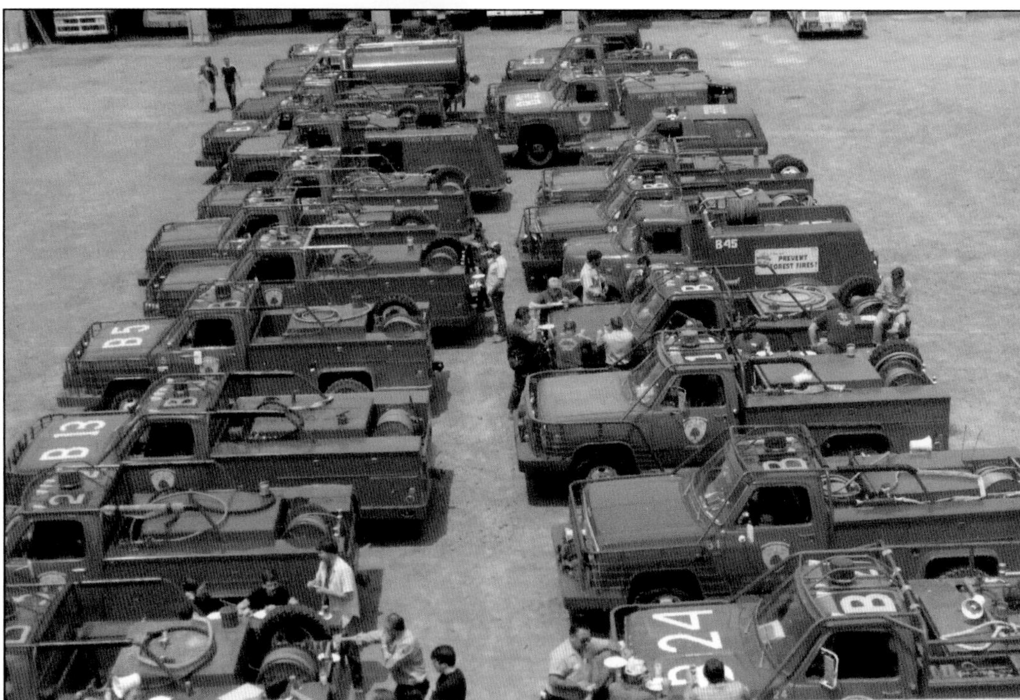

The New Jersey Forest Fire Service has relied on mechanized equipment to help combat fires since the 1930s. Once four-wheel drive vehicles were developed, off-road engines proved to be a great asset in rodeoing fast-spreading wildfires where the engine could encircle the fire and cool it with water. This 1985 photograph shows a fleet of central New Jersey engines assembled for a firefighter training session. (New Jersey Forest Fire Service.)

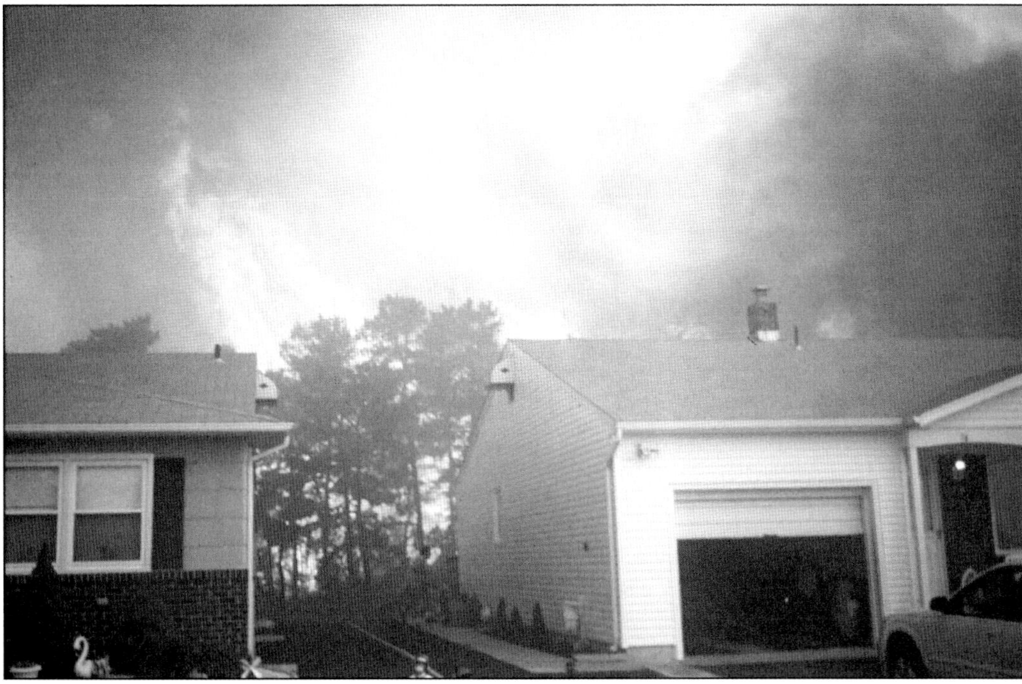

The New Jersey Forest Fire Service combats about 1,600 wildfires on an annual basis across the state. This photograph shows a wildfire threatening homes during the Wrangle Brook fire in Ocean County in 1997. The weekend of April 20–22, 1963, serves as a benchmark for all northeastern states for wildfire potential in this region of the United States. On these three days, under extreme fire conditions, nearly 190,000 acres of forest were destroyed, seven people lost their lives, 2,500 were evacuated, and 1,000 were left homeless. (New Jersey Forest Fire Service.)

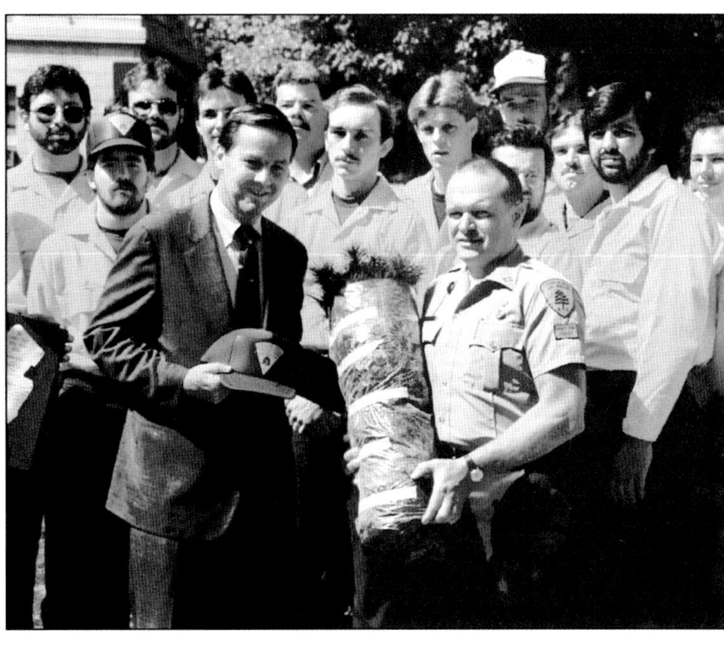

In 1988, the New Jersey Forest Fire Service was awarded the Jersey Pride Award from Gov. Thomas Kean for their brave work in combatting fires in state, as well as in and around Yellowstone National Park that year. State fire warden David B. Harrison accepted the award. Harrison was a member of the Northeast Forest Fire Supervisors for 18 years, past chairperson, and also a recipient of the NFFS Fire Supervisors Award in 1989. (New Jersey Forest Fire Service.)

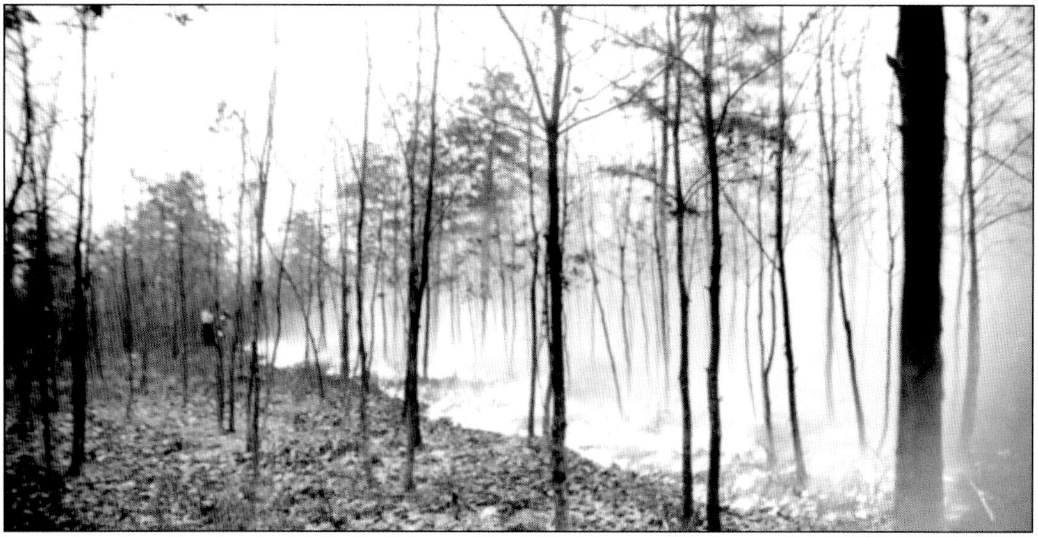

Prescribed fire has been used in New Jersey to meet fire management and public safety objectives since the 1910s, and one of the first documented uses was to protect cranberry bogs from wildfire. The New Jersey Forest Fire Service first used prescribed fire in 1928, and this 1937 photograph shows some of the first documented use of prescribed fire on state-owned land where 100-foot safety strips were created along roads at the Lebanon Experimental Forest. Today, the New Jersey Forest Fire Service targets treating 20,000 acres of open space with prescribed fire annually. (New Jersey Forest Fire Service.)

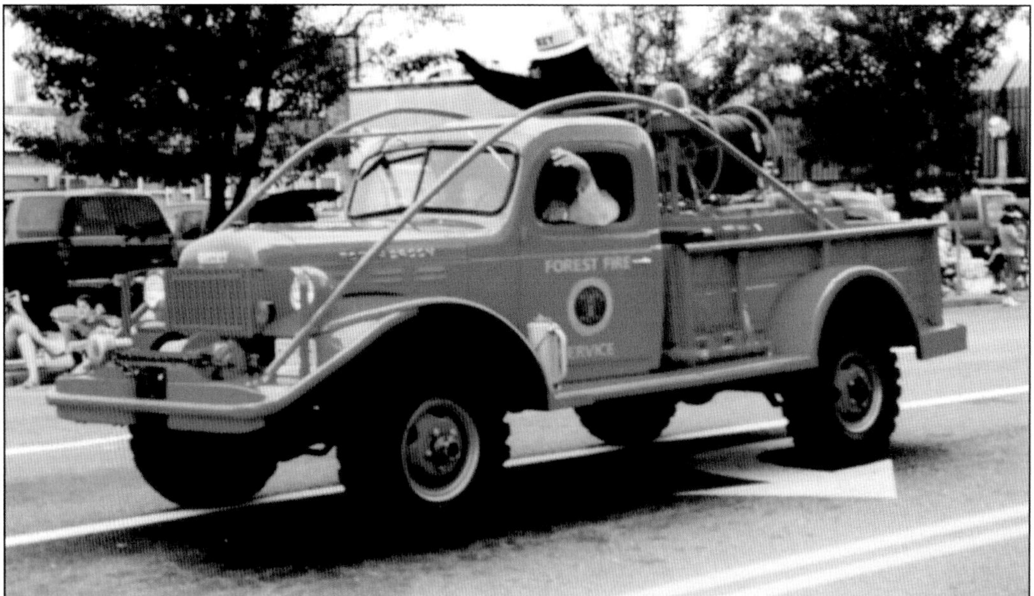

In 2006, the New Jersey Forest Fire Service celebrated its centennial, and this 1969 WM model Dodge Power Wagon was refurbished specifically for use at parades and events to help commemorate this great milestone. This Dodge Power Wagon represents the backbone of the fire service as the workhorse that helped wildland firefighters battle fires and perform their duties on a daily basis from the 1950s through today. Even though today's trucks are not as tough as this one, off-road engines are an integral part of fulfilling the fire management mission. (New Jersey Forest Fire Service.)

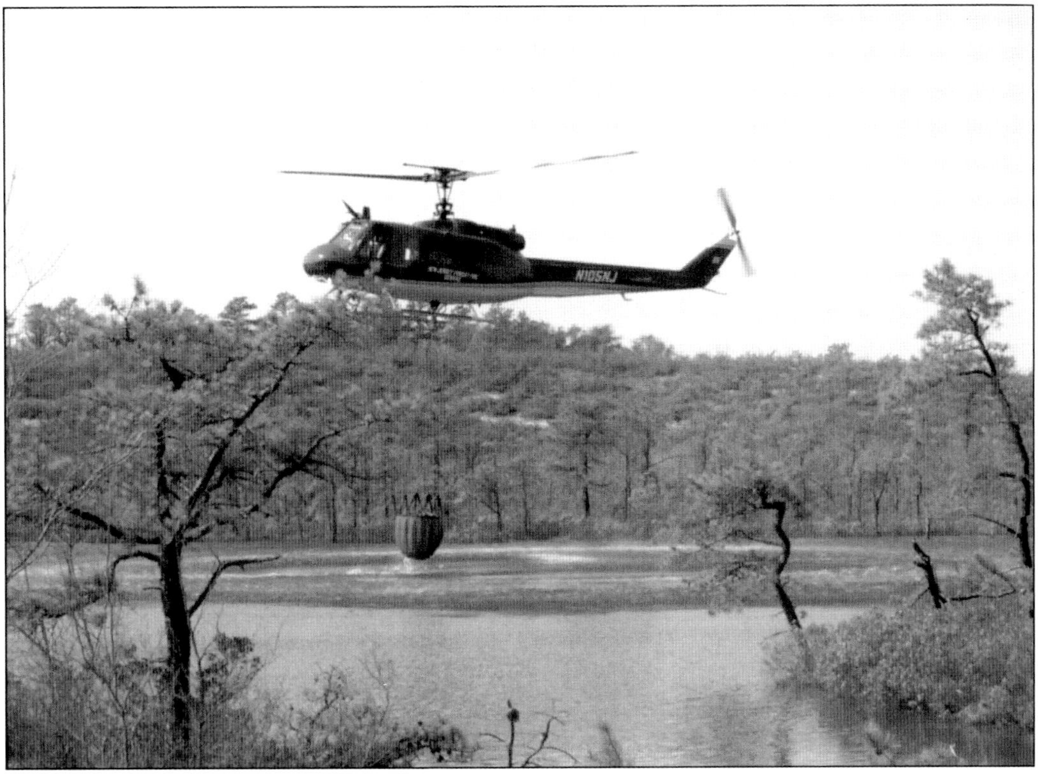

The New Jersey Forest Fire Service has used aircraft to combat fires, aid in their detection, and perform aerial reconnaissance and surveillance since 1927. Over the years, aircraft has been state owned, contracted from private aviation vendors, or on loan from the US Forest Service through the Federal Excess Personal Property (FEPP) program. This photograph shows an FEPP Bell UH1H helicopter picking up a load of water near the Warren Grove Bombing Range. (New Jersey Forest Fire Service.)

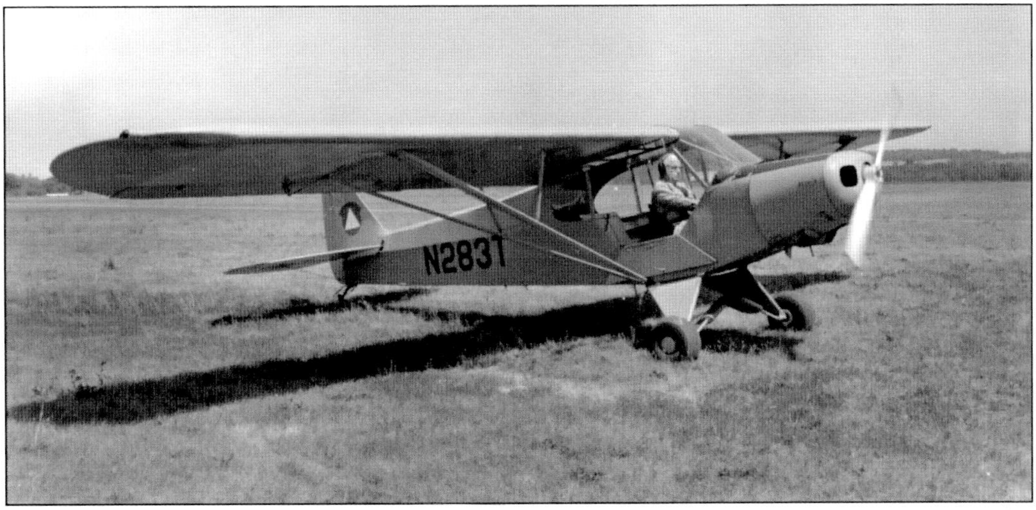

By the 1950s, the Maryland Department of Natural Resources Forest Service began to employ airplanes to spot forest fires. Here, forest supervisor Herman Toms prepares to take off for a forestry demonstration at the Frederick Fair on August 30, 1962. (Maryland Forest Service.)

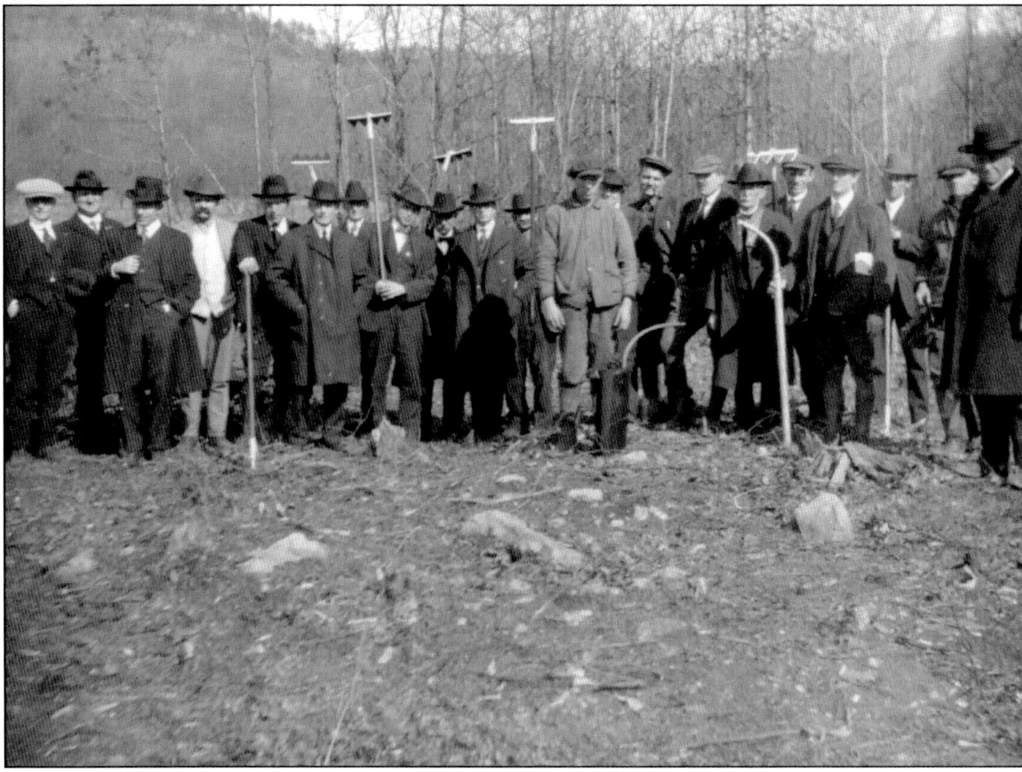

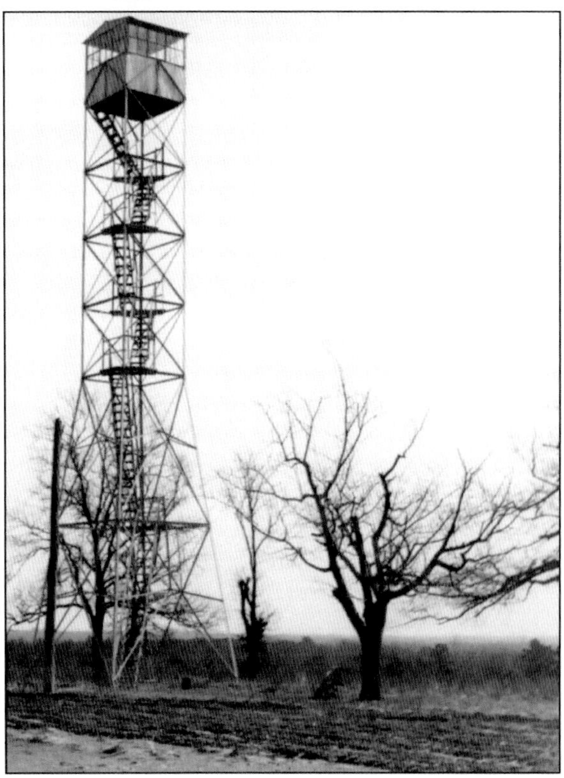

The Maryland Department of Natural Resources Forest Service was founded in 1906, as the State Board of Forestry, after a generous land grant from Robert and John Work Garrett. The first state forester, Frederick W. Besley, understood the importance of fighting forest fires and created a network of volunteer forest wardens. Pictured above are the forest wardens of Frederick and Washington Counties in 1922. (Maryland Forest Service.)

Before modern communication networks, the Maryland Forest Service depended on fire towers to spot and locate forest fires. Most of the 34 fire towers across the state were constructed by the Civilian Conservation Corps in the early 1930s. There were several towers constructed earlier, such as the one at Long Hill, pictured at left in 1923. Long Hill later became the central communication center for the Maryland Department of Natural Resources. (Maryland Forest Service.)

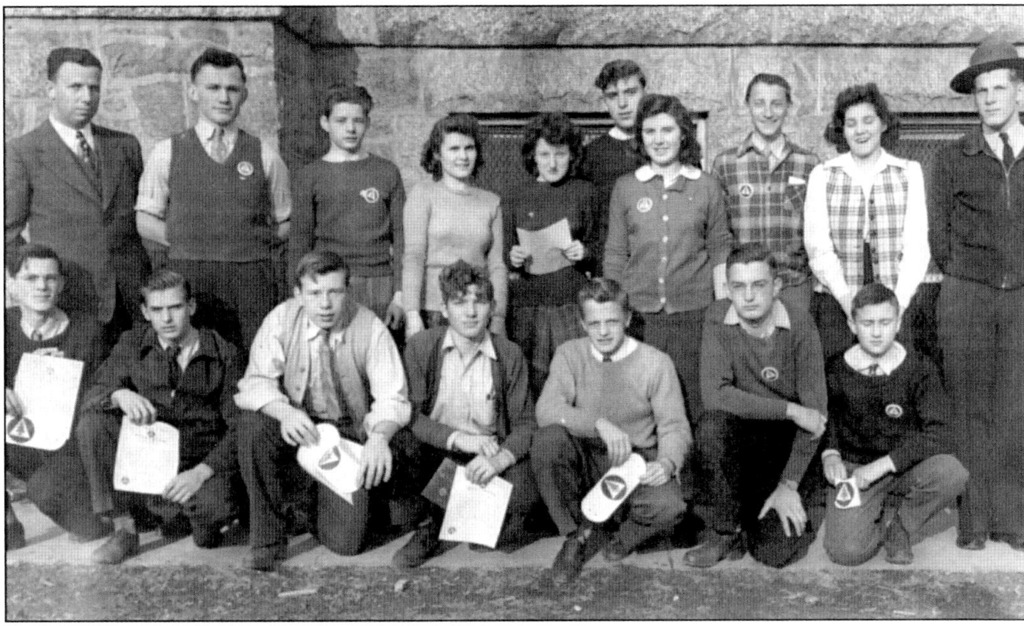

During World War II, many of the Maryland Department of Natural Resources Forest Service field staff were pressed into military service, forcing the agency to find creative means to fight forest fires. As part of their civil defense training, high school students, like the ones pictured above, were mustered into forest firefighting brigades. (Maryland Forest Service.)

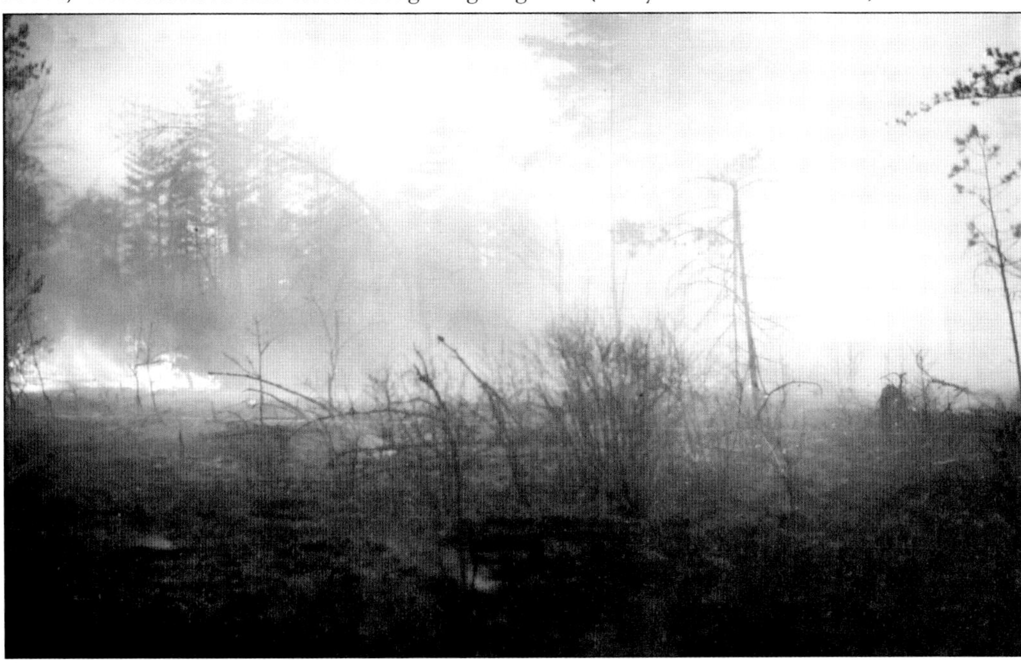

Early-20th-century foresters viewed fires as the ultimate threat. Over 30,000 acres burned in Maryland annually, destroying thousands of dollars of property. Here, a fire burns through Beltsville, in Prince George's County, ultimately burning 1,000 acres in April 1926. Unlike in 1926, hundreds of homes are now located in the area where this fire burned—just one example of Maryland's prominent wildland-urban interface. (Maryland Forest Service.)

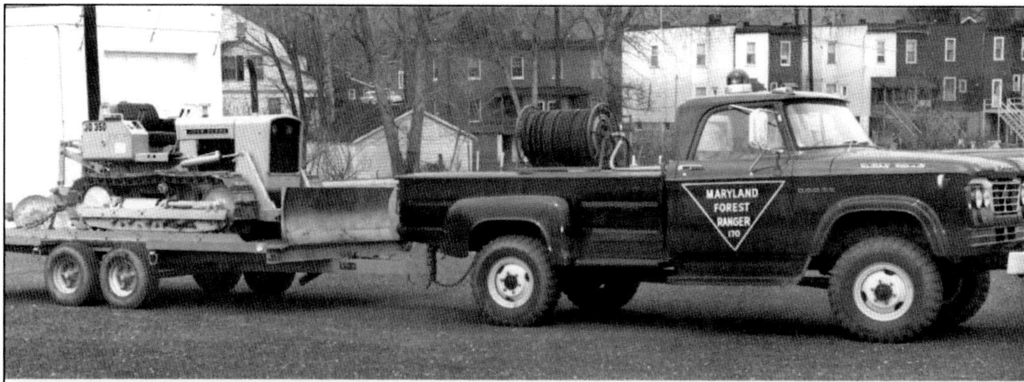

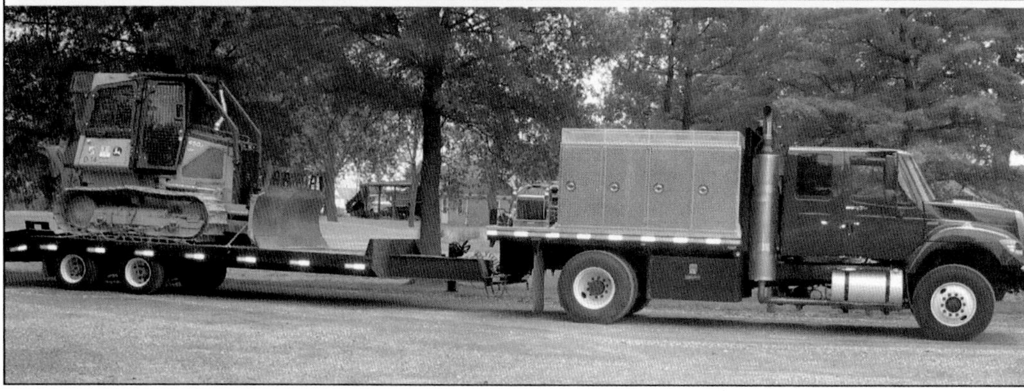

Over time, the Maryland Forest Service firefighting fleet has increased in capability, sophistication, and safety. The typical unit from 1962, pictured above, had many limitations and afforded no protection for the tractor operator. Modern tractor-plow units have mesh-enclosed cabs, roll-over protection, and other specialized firefighting modifications. The tractor unit above is towed by a Type 4 engine that has ample room for water, hose, tools, specialty equipment, and crew members. (Maryland Forest Service.)

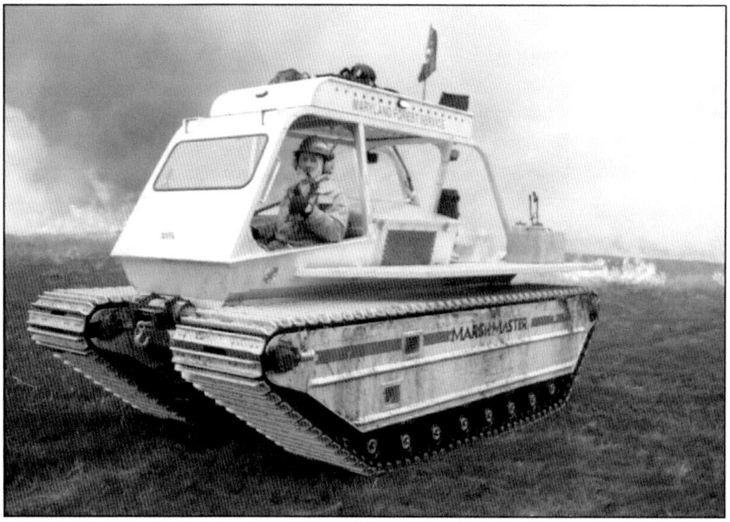

Maryland is home to thousands of acres of salt marsh along the Chesapeake Bay. Marsh is a unique and volatile fuel type, which presents definite challenges to firefighting efforts. To help overcome these obstacles, the Maryland Forest Service uses specialty equipment, such as this Marsh Master amphibious firefighting apparatus. (Maryland Forest Service.)

In 1921, the Ohio Legislature passed the Silver Bill, which provided for the appointment of a state forester and funding for land acquisition and for the organization a forest fire control system. By the end of 1921, the fire control program consisted of one chief warden and 96 district and deputy wardens and protected approximately 600,000 acres in southern and eastern Ohio. Prior to 1922, it was estimated 50,000 to 70,000 acres burned annually in Ohio. (Ohio DNR Division of Forestry.)

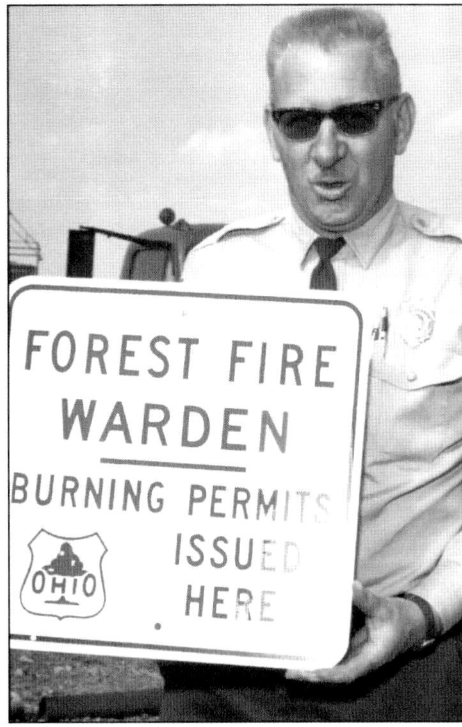

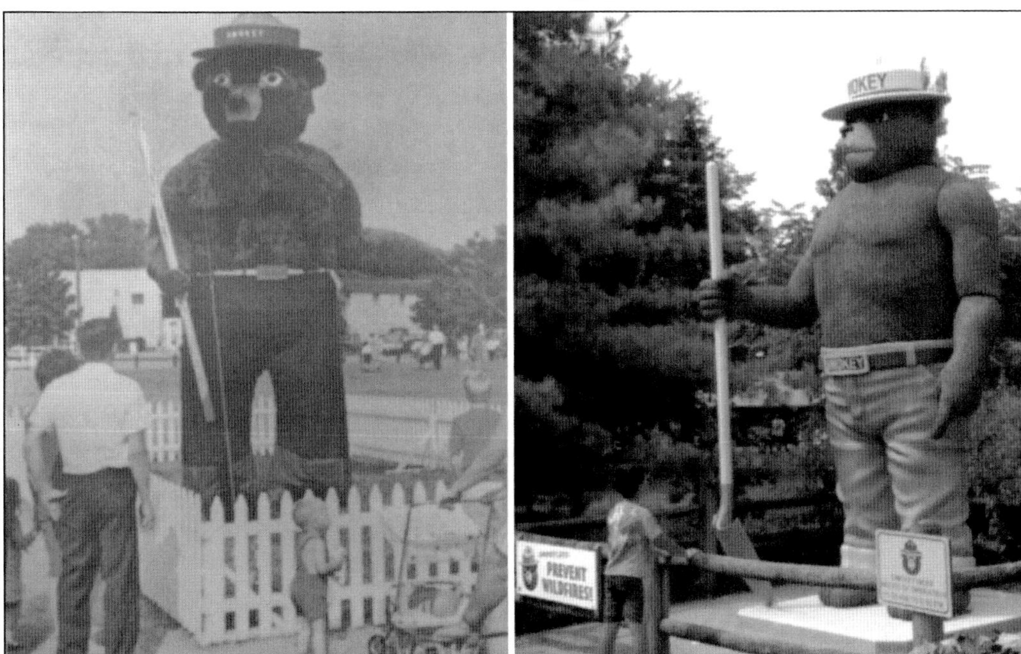

Smokey Bear first appeared at the Ohio State Fair in 1959 when a 14-foot-tall animated likeness was created by forestry carpenters from wood, chicken wire, and fiberglass. This Smokey greeted guests at the fair for over 50 years until 2015, when a brand new animated Smokey was unveiled. During his time at the Ohio State Fair, Smokey Bear has reached millions of Ohioans with his wildfire prevention message. (Ohio DNR Division of Forestry.)

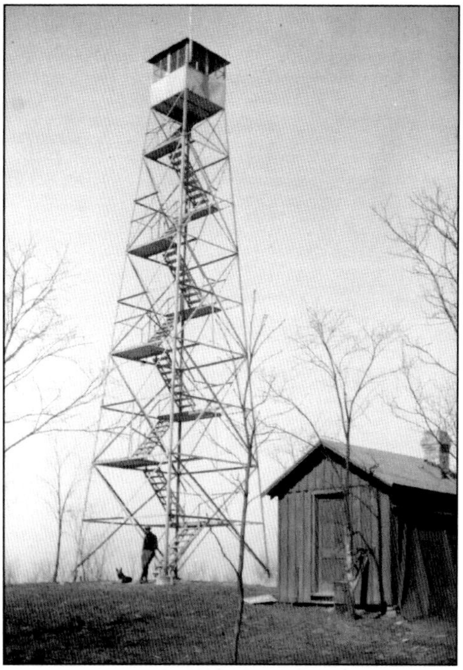

The first fire tower in Ohio was built in 1924 on Copperhead Hill in Shawnee State Forest, Scioto County. A total of 39 fire towers were erected in Ohio, and the last one was constructed in 1970. The last active fire tower, Green Ridge Tower in Pike County, was shut down in 1978. Although no longer used for fire detection, seven fire towers remain standing on state forestland and are frequently climbed by hikers who wish for a beautiful view. Pictured is W.M. Debolt at the Scioto Trail Fire Tower, Ross County. (Ohio DNR Division of Forestry.)

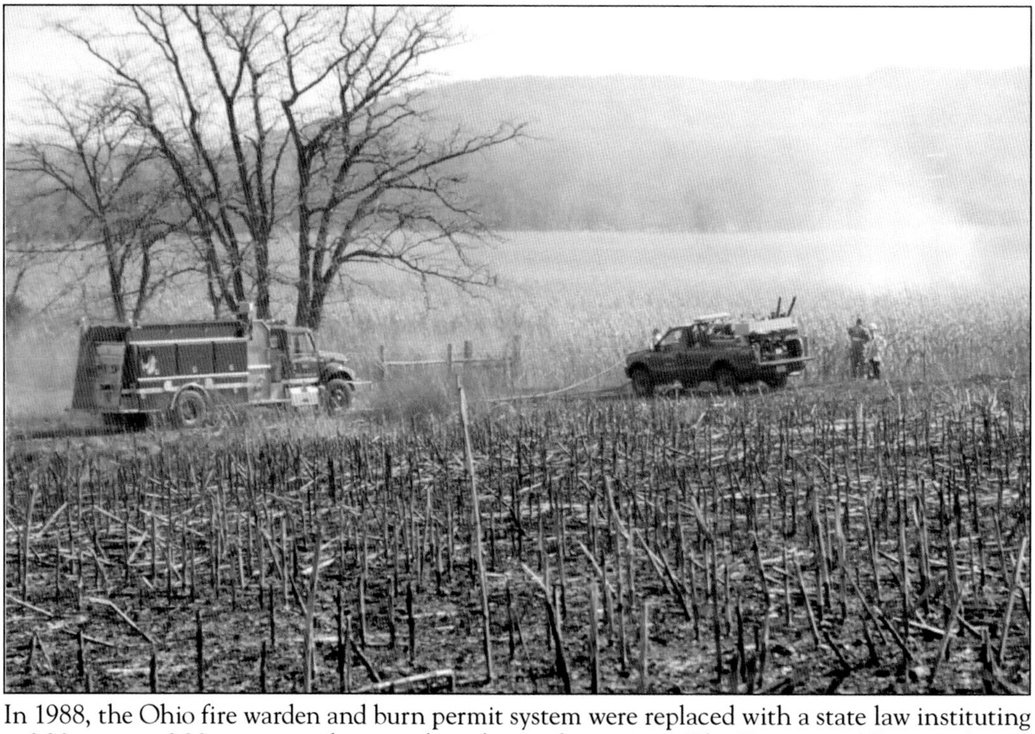

In 1988, the Ohio fire warden and burn permit system were replaced with a state law instituting a 6:00 a.m. to 6:00 p.m. open-burning ban during fire season. The Division of Forestry began paying rural fire departments for the suppression of wildfires in southern and eastern Ohio. In 1992, the agency closed five fire control field stations and eliminated the central fabrication shop and the entire aviation program. Pictured is a local fire department responding to a field fire in Adams County. (Ohio DNR Division of Forestry.)

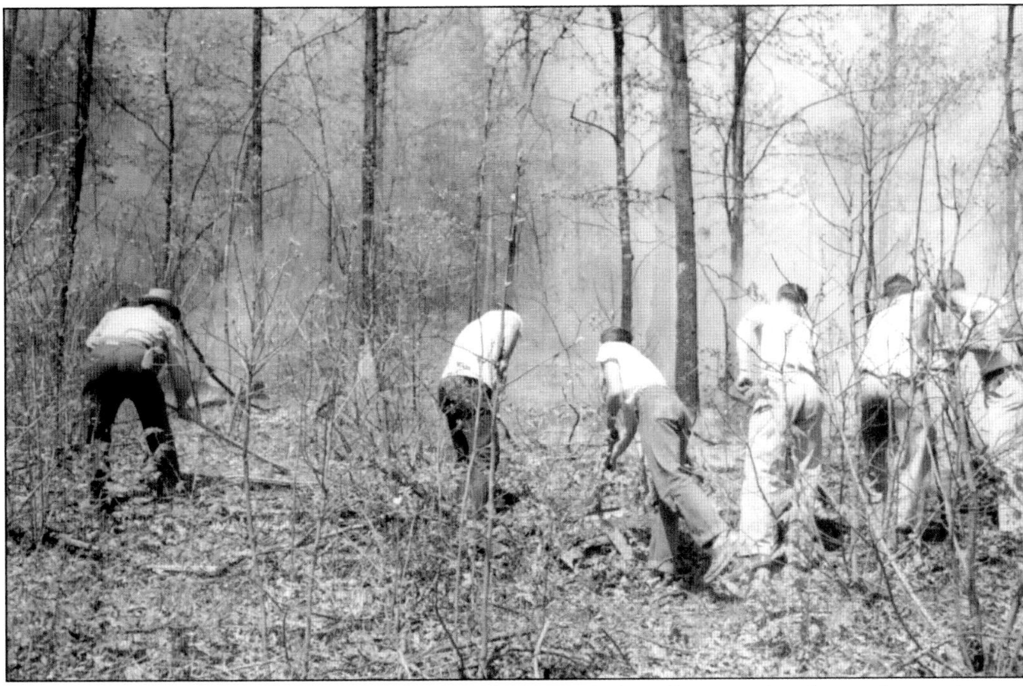

During the 1930s, the Civilian Conservation Corps contributed tremendously to reforestation and fire control in Ohio. Twenty-six CCC camps, each accommodating 200 men, were established in Ohio, and 14 of those were located on state forests or parks. CCC members built roads and fire trail networks, planted millions of trees, and fought numerous fires. They constructed 12 fire towers and a forest fire control telephone network, which had more miles of line than some local telephone companies at the time. (Ohio DNR Division of Forestry.)

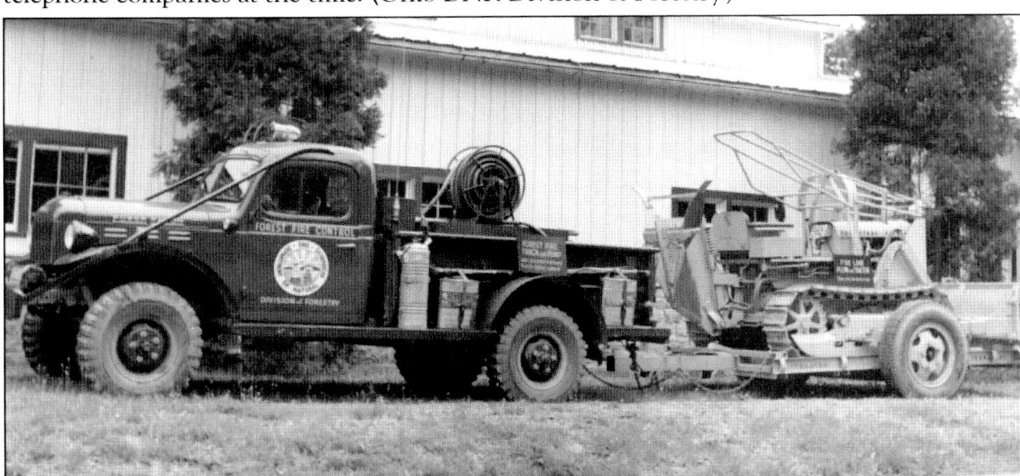

By the 1950s, the formerly vast CCC workforce had dwindled, leaving the Ohio Division of Forestry short on manpower to fight wildfires. In response to this, efforts to mechanize the forest fire suppression force began. Engines, dozers, fire plows, and aircraft were acquired, designed, and fabricated by the Division of Forestry, changing how fires were fought in Ohio. Pictured are a Division of Forestry Dodge Power Wagon and a Terratrac dozer outfitted for wildfire suppression. (Ohio DNR Division of Forestry.)

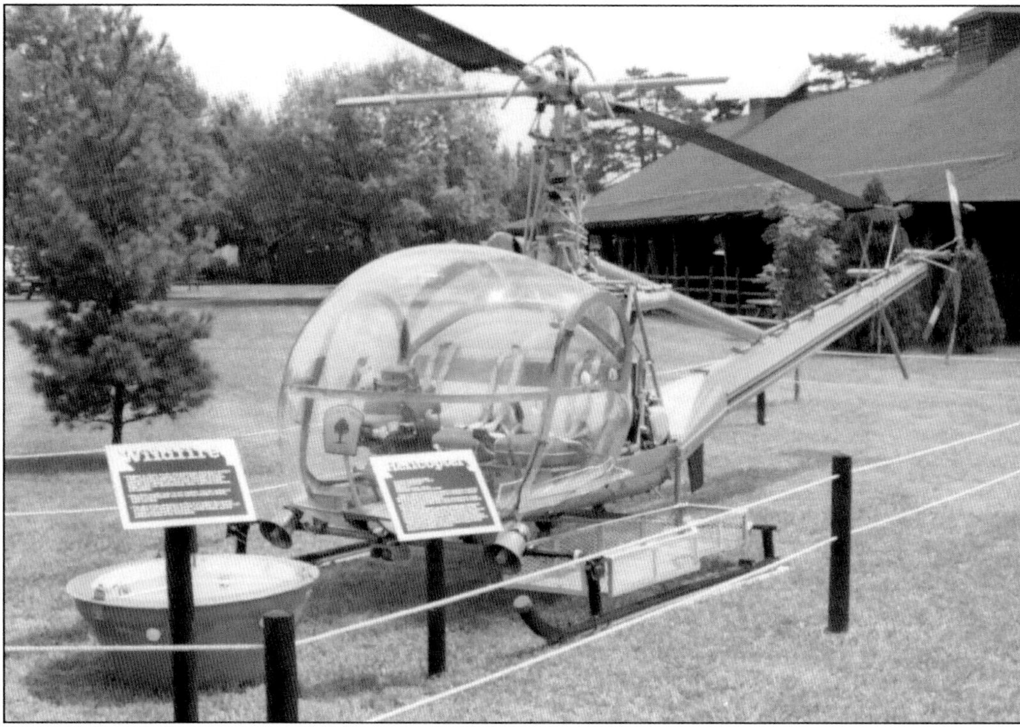

Aircraft, including the Hiller helicopter, was a very important part of the Ohio fire program from the 1960s through the 1980s. Aerial detection was so effective that it eventually caused the phasing out of fire towers. During the peak of the aviation program, the Ohio Division of Forestry employed seven full-time pilots, supplemented with contractor pilots, and operated nine airplanes and five helicopters. The aviation program declined by late 1980s as costs escalated and equipment aged. (Ohio DNR Division of Forestry.)

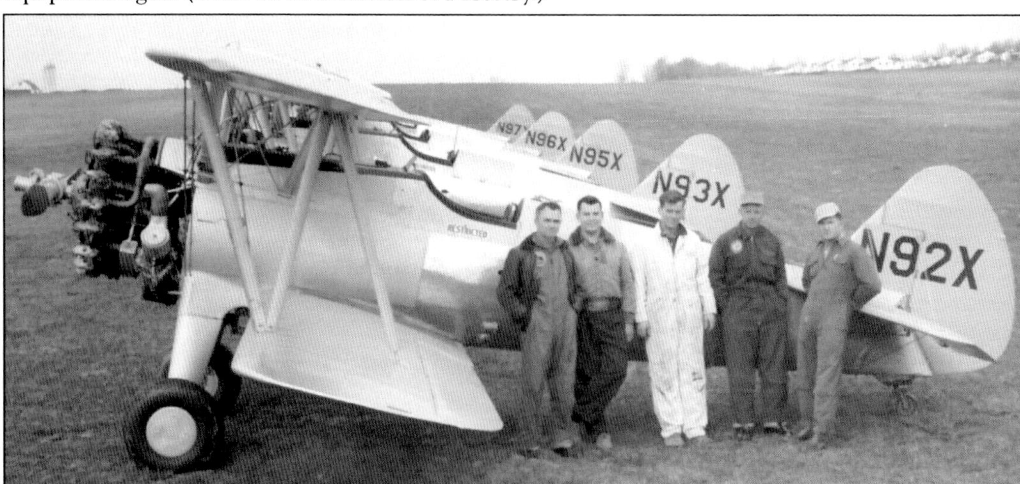

This 1961 photograph shows Pennsylvania's air tanker fleet that was utilized to combat wildfires throughout Pennsylvania. The pilots are, from left to right, T.R. Coats, L.R. "Buddy" Lewis, John P. Stoltzfus, Wayne North, and John Klein. Over the years, Pennsylvania's aviation fleet has included B-26s, P-38s, Stearman, Bell helicopters, TBMs, Sikorsky S-55, Beechcraft, C-47, PV-2s, F-7F Tigercat, Dromaders, and Airtractors. (Pennsylvania DCNR Division of Fire Protection.)

This is the Whittimore Fire Tower and helibase at Pennsylvania's Elk State Forest in the 1960s. Currently, the state of Pennsylvania has 45 fire towers, and most of them are still utilized in fire season. Recently, the Pennsylvania Division of Forest Fire Protection received funding to renovate and rebuild many of the aging structures. (Pennsylvania DCNR Division of Fire Protection.)

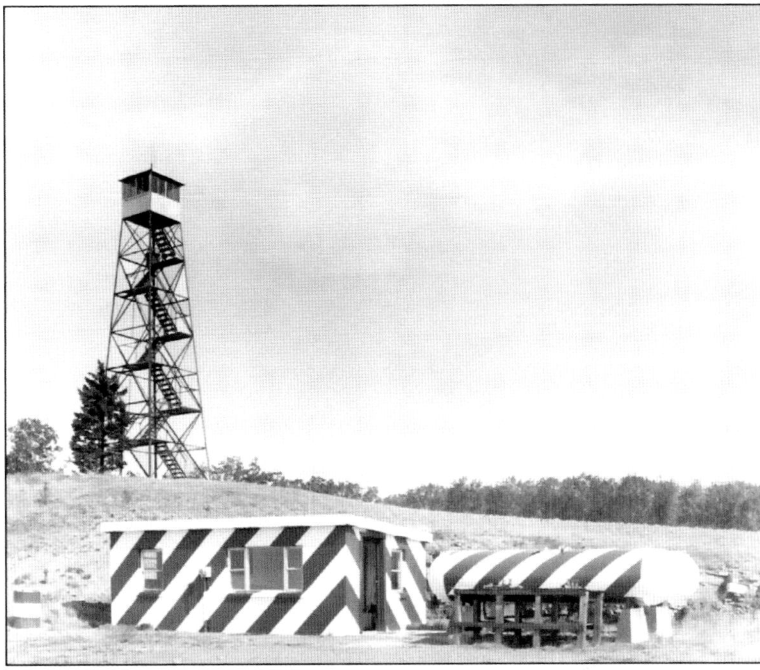

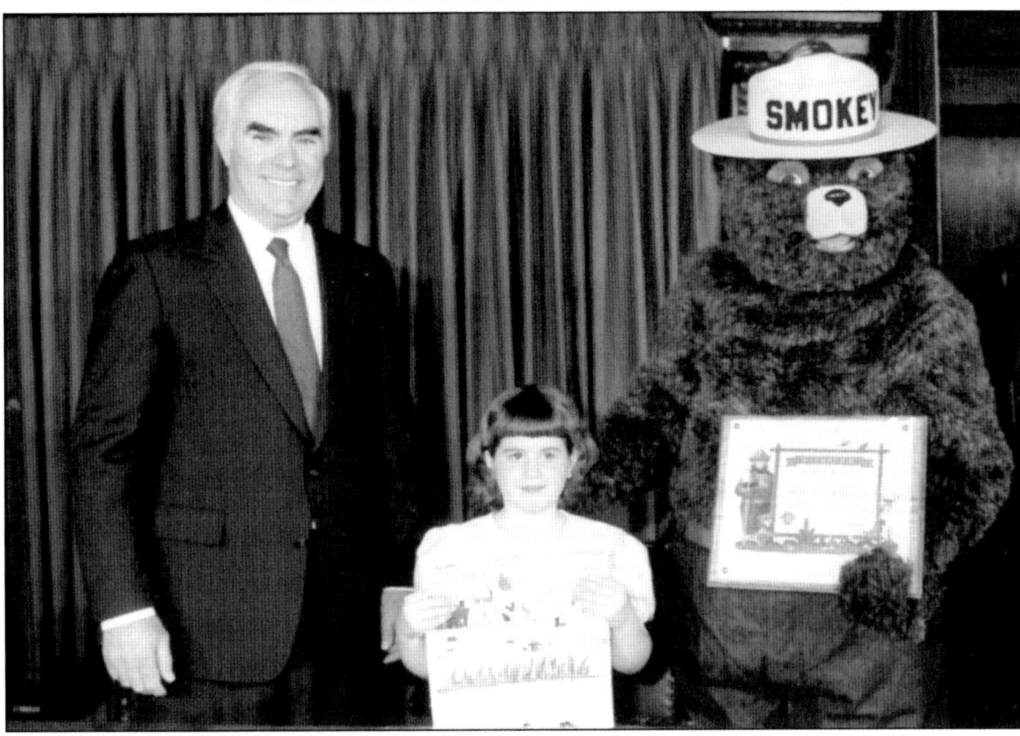

Pictured above are Pennsylvania governor Robert Casey (left) and Smokey with the winner of a Smokey Bear poster contest. Wildfire prevention has always been a priority in Pennsylvania. Smokey makes many appearances to schools and other public events to raise awareness and encourage people to become active proponents of wildfire prevention. (Pennsylvania DCNR Division of Fire Protection.)

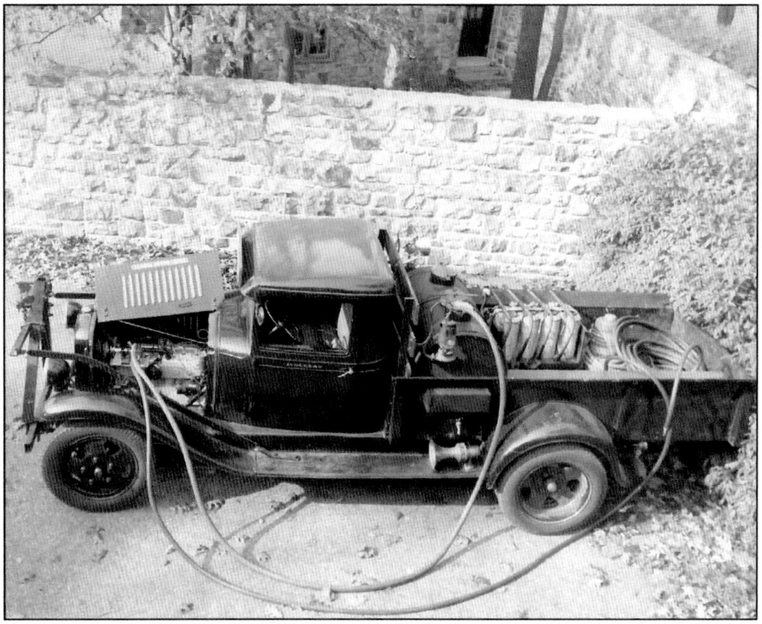

The Pennsylvania Forest Fire Warden program was established in 1915 and has been a strong asset for fighting wildfires. The program always utilized the best equipment for their time period, which included vehicles like this antique truck. This picture shows the ingenuity and resourcefulness of the wardens to use available equipment to the best of their ability. (Pennsylvania DCNR Division of Fire Protection.)

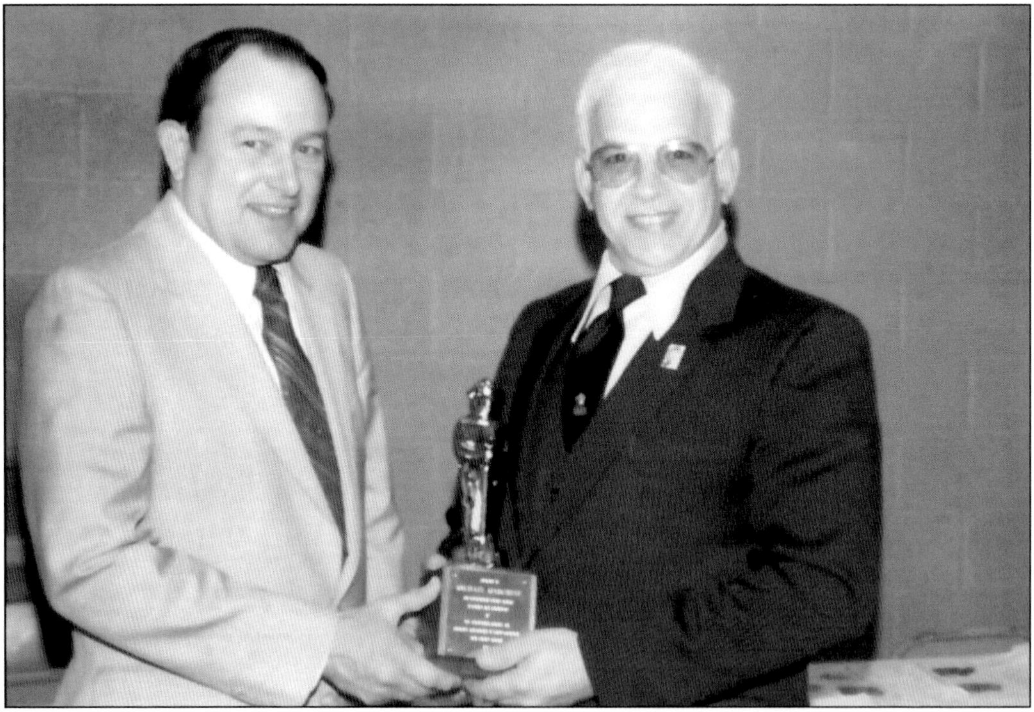

Mike Marchese (right) is receiving the Silver Smokey Award from Tom Schenart (left). Marchese played an integral part in the issuance of the Smokey Bear postage stamp to commemorate Smokey's 40th birthday. More recently, in 2013, the Wardens Helping in Prevention (WHIP) program received the Bronze Smokey Award for their efforts in wildfire prevention. (Pennsylvania DCNR Division of Fire Protection.)

The Pennsylvania Division of Forest Fire Protection provides resources to assist in other states when wildfire conditions exceed their capabilities. The picture at right was taken in 2015 in Washington and depicts a newer Type 6 engine. On average, Pennsylvania mobilizes 100–200 firefighters per year. (Pennsylvania DCNR Division of Fire Protection.)

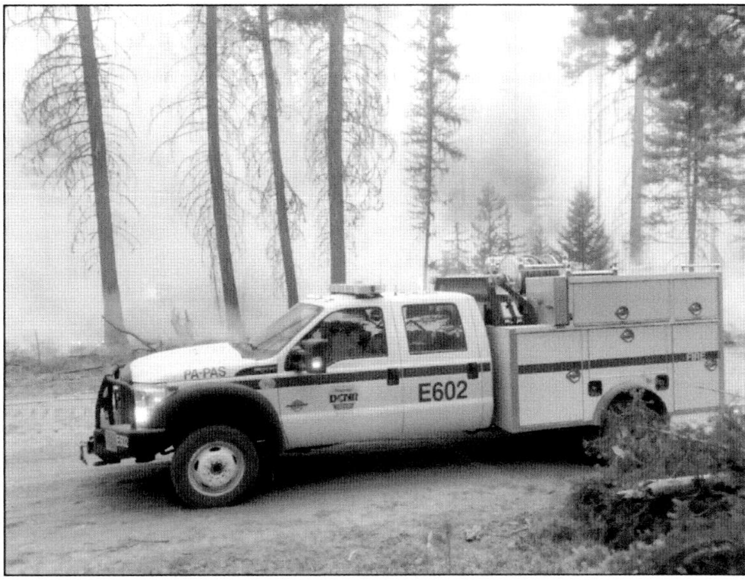

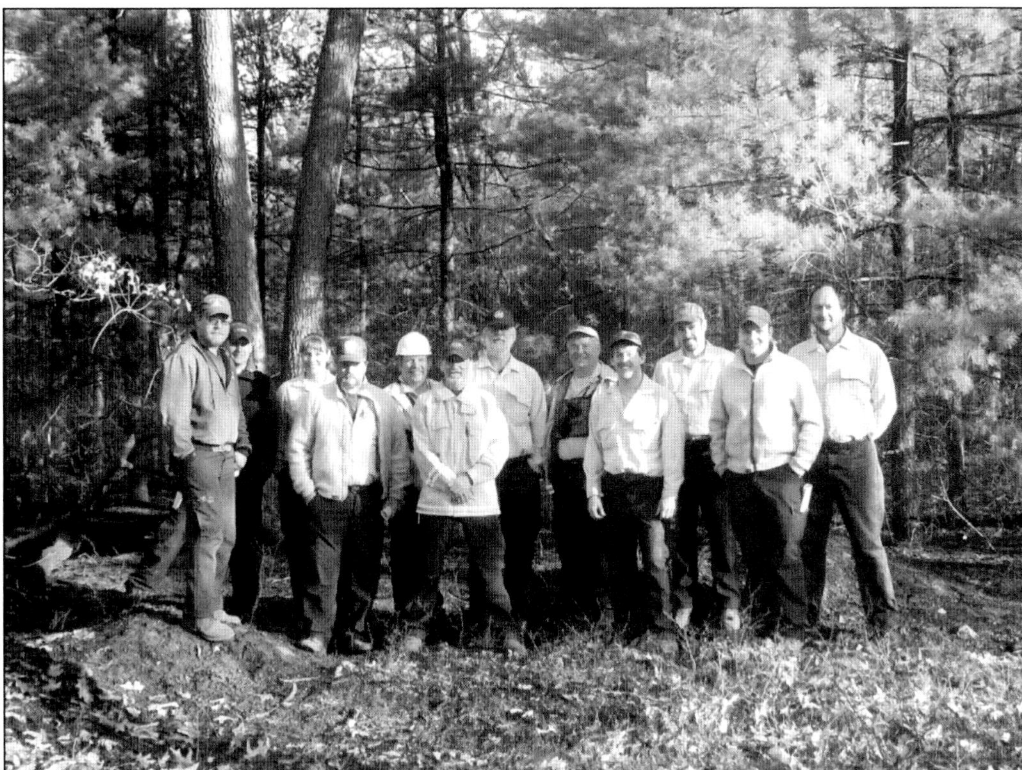

This is Pennsylvania's first Type 3 Incident Management Team, which was deployed in May 2006. They managed the Okame Road fire in the Loyalsock State Forest. The fire was large at 602 acres, and it was a great learning tool to propel the further development of Pennsylvania's Type 3 teams. Pennsylvania currently has two teams in rotation, with many of the team members participating on national teams. (Pennsylvania DCNR Division of Fire Protection.)

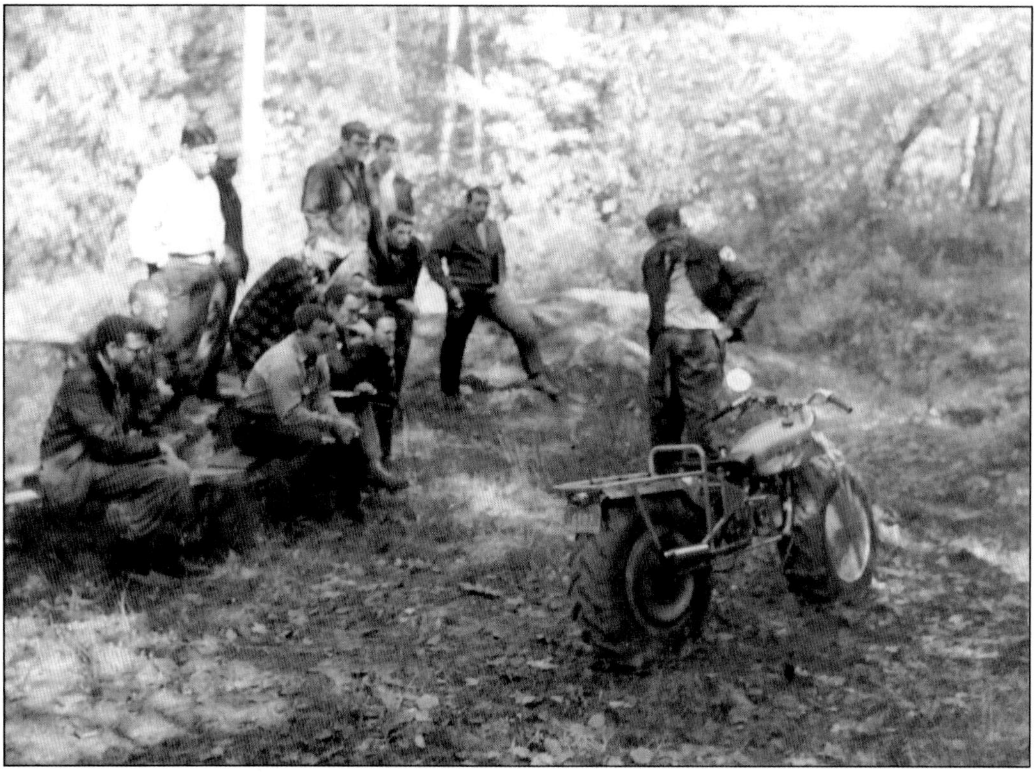

Railroads were a major cause of wildfires throughout West Virginia until the 1970s, when railroads installed better-quality and more efficient spark arrestors on their train engines. In 1962, the Division of Forestry acquired its first ATV to patrol tracks in inaccessible areas where railroad-caused wildfire occurrence was high. In this photograph, the safe use of the Gama goat is shared among forestry firefighters. (West Virginia Division of Forestry.)

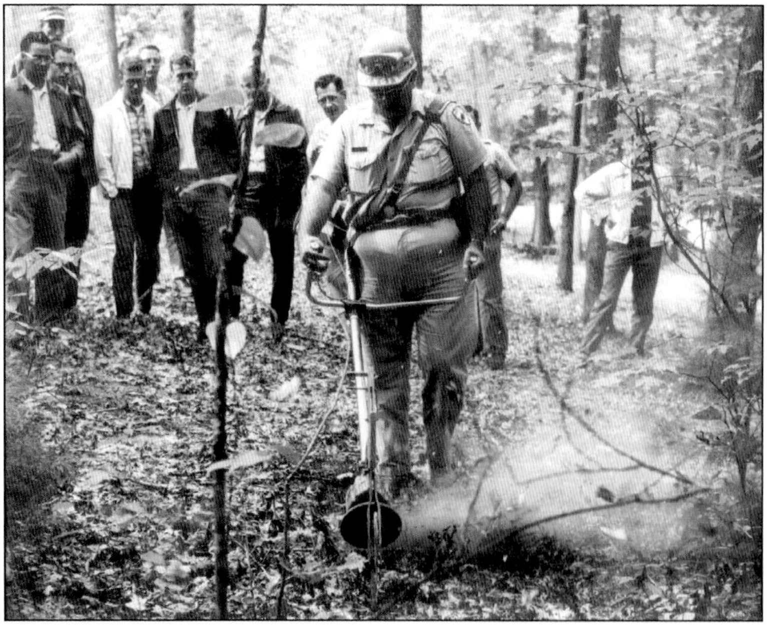

In the late 1970s, the West Virginia Division of Forestry experimented with its first mechanized hand line construction tool. It was called a flail trencher with a head composed of chain-link sections that rotated and threw leaf litter and debris to the side. After field testing and use, the flail trencher was deemed ineffective and too hazardous to the operator. (West Virginia Division of Forestry.)

In 1922, the West Virginia Division of Forestry purchased its first steel structure fire towers, and the initial one was constructed on Backbone Mountain in Tucker County. At one time, West Virginia utilized over 70 fire lookout towers. In 1987, West Virginia's fire towers were no longer utilized for wildfire detection, and most were dismantled or sold. Today, remaining lookout towers are used for radio antenna mounts to aid with communications. (West Virginia Division of Forestry.)

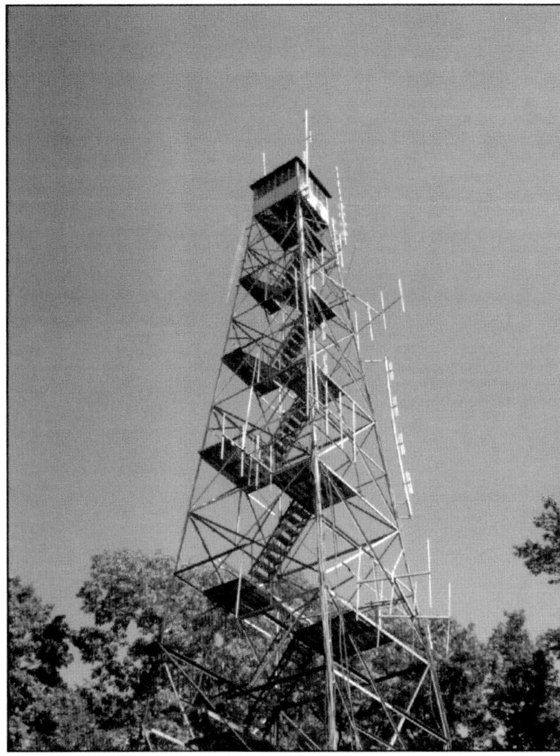

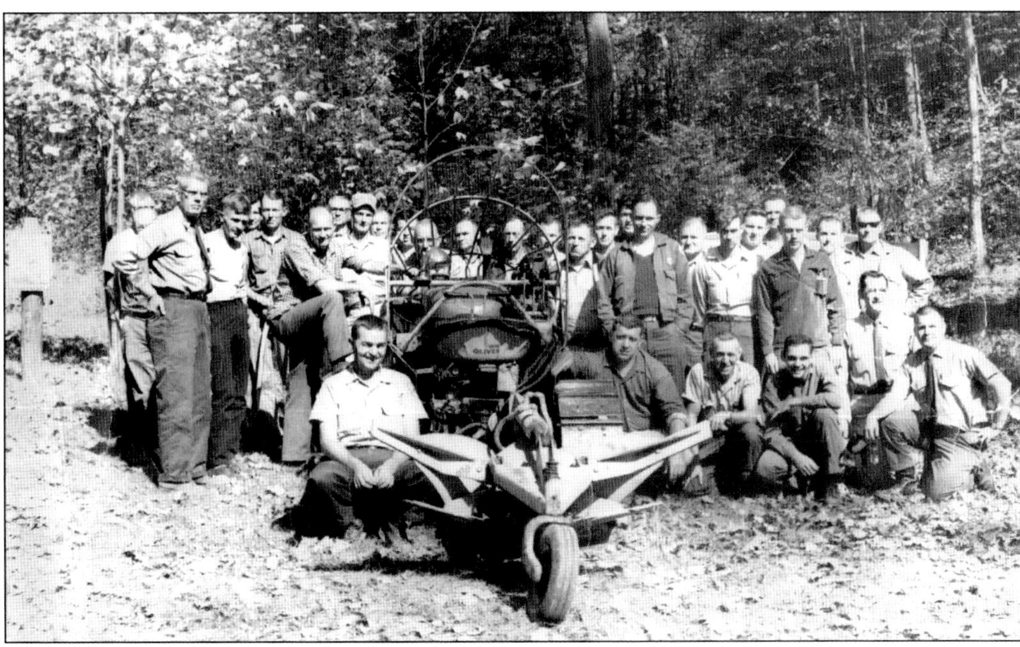

In 1963, the West Virginia Division of Forestry acquired its first piece of heavy equipment, an Oliver tractor equipped with a brush guard and fire plow. This photograph captures the graduates of West Virginia's annual wildfire training school surrounding the Oliver tractor. The training school graduates included temporary employees that aided in the suppression of wildfires during the state's fire seasons. (West Virginia Division of Forestry.)

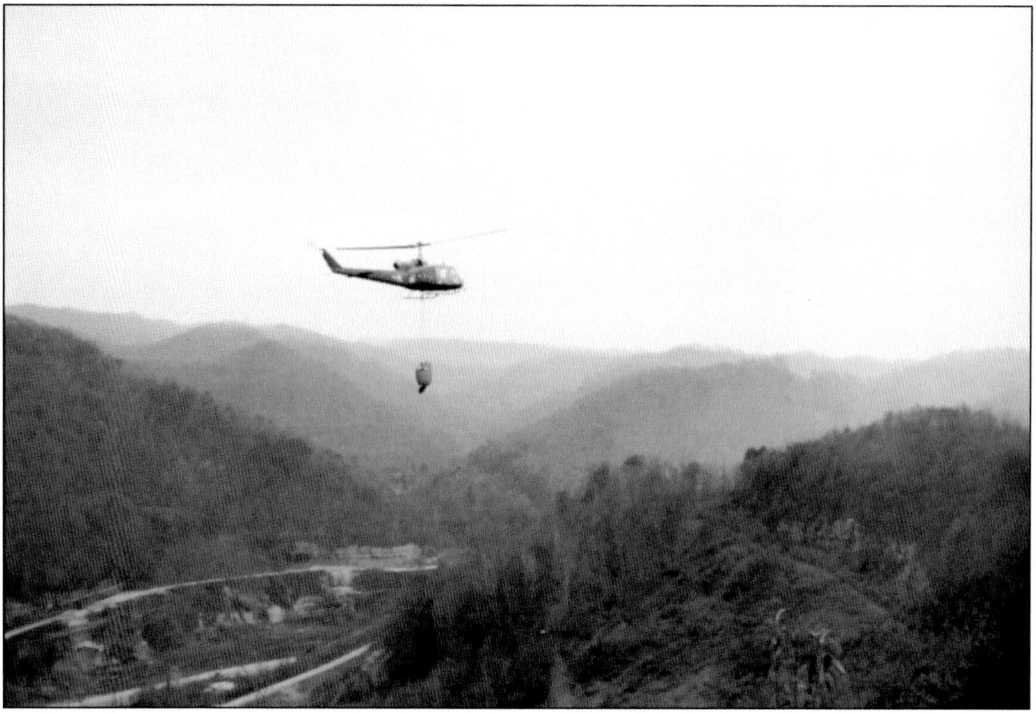

In 1999, the West Virginia Division of Forestry initiated its first contract for aerial firefighting resources with a single-engine air tanker (SEAT). In 2004, the SEAT was replaced with a helicopter. This contract Huey helicopter was equipped with a 300-gallon bucket and was a valuable tool in supporting West Virginia's wildland firefighters in containing fires in the state's rugged terrain as visible in this photograph. This helicopter served the US Army in Vietnam and, in addition to firefighting, has appeared in several action movies. (West Virginia Division of Forestry.)

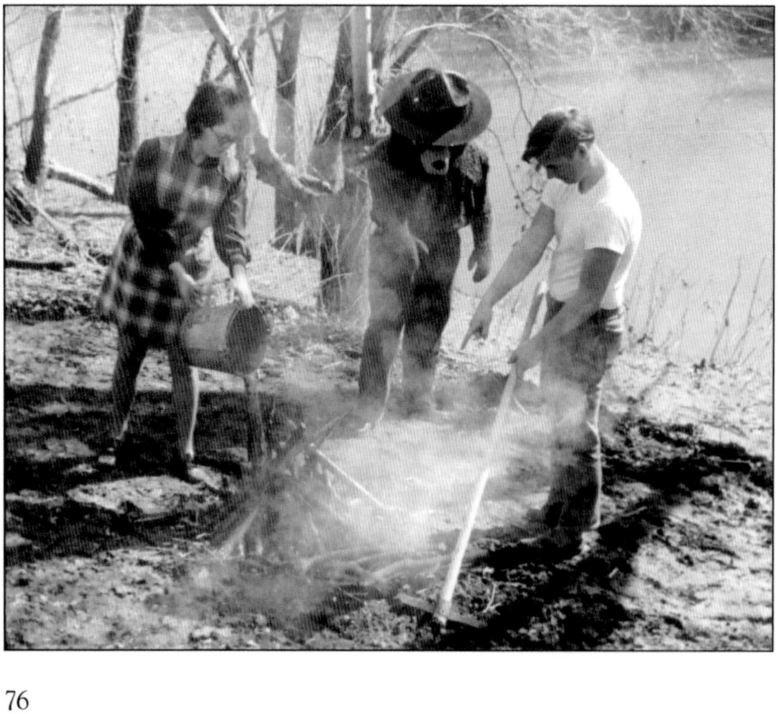

Smokey Bear has been used to convey the wildfire prevention message throughout West Virginia since his inception in 1944. In this 1950s photograph, one of the first Smokey Bear characters teaches the method to properly extinguish campfires as part of West Virginia's fire prevention campaign. (West Virginia Division of Forestry.)

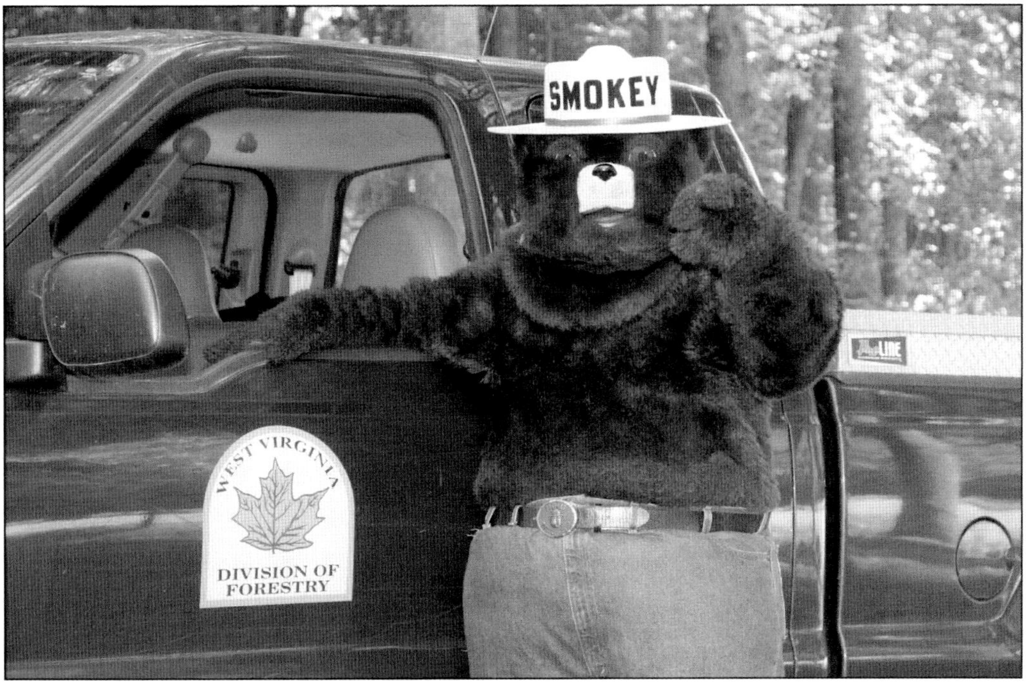

Today, Smokey Bear continues to spread his wildfire prevention message at schools, fairs, and festivals with the help of West Virginia Division of Forestry staff. He makes classroom appearances and meets over 10,000 students each year and greets over 25,000 visitors at fairs and festivals across West Virginia. (West Virginia Division of Forestry.)

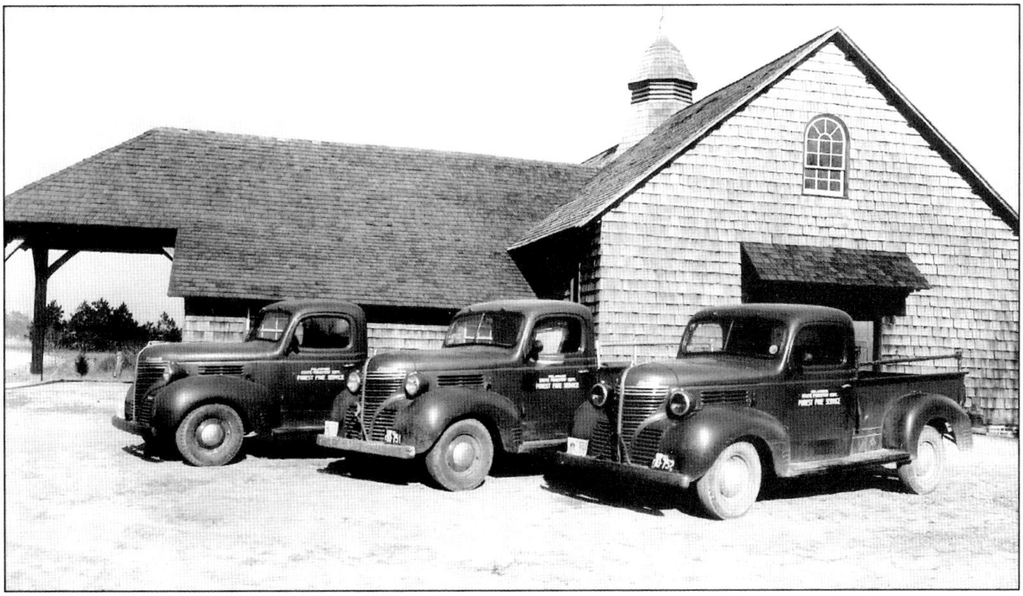

This 1945 photograph shows three Dodge trucks near the historic barn at Redden State Forest; the trucks were acquired as a gift from the US Forest Service in 1942 and used as fire patrol trucks by the Delaware Forest Service. The trucks are lined up on the south side of the barn at Redden. The historic barn is part of a complex of buildings, including the Forest Lodge and the forest ranger station, that are listed in the National Register of Historic Places. (Delaware Forest Service.)

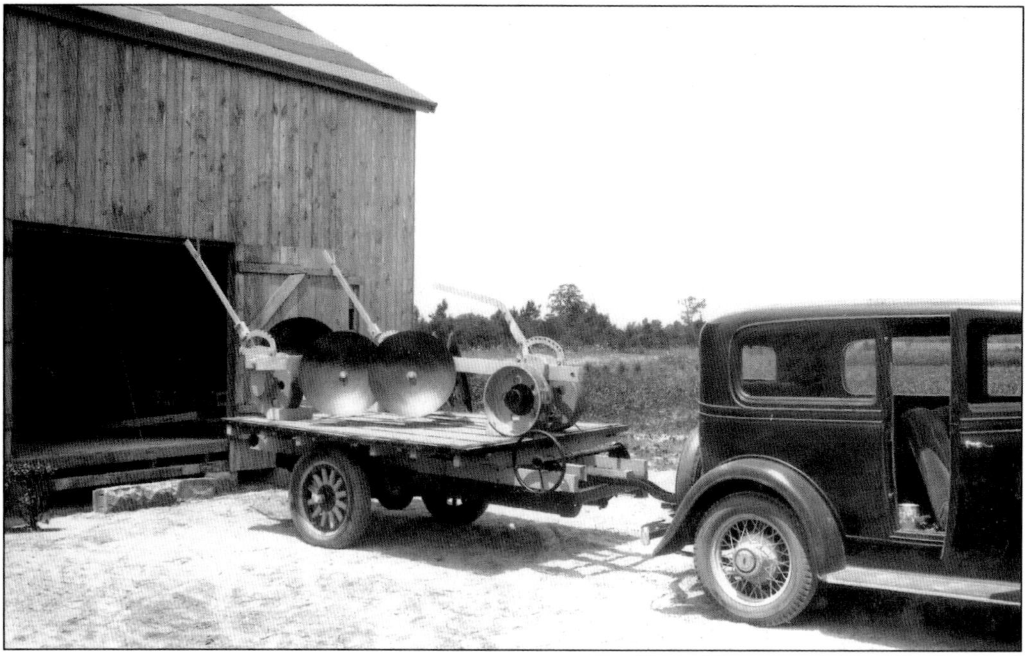

This 1933 photograph shows the Hester fire break plow purchased by the Delaware Forestry Department in 1932. The trailer was built during the winter of 1932–1933 from a Chevrolet truck purchased for $20. The trailer and special drawbar were designed by Delaware's first state forester William S. Taber and C.D. Goodrich. The drawbar has a double set of take-up springs. (Delaware Forest Service.)

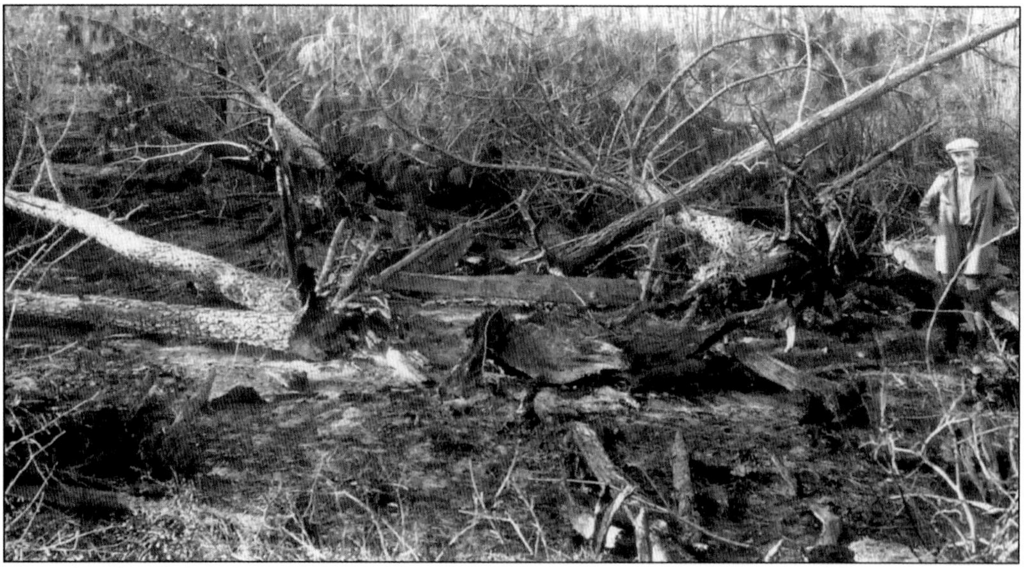

Large fires burned in the Great Cypress Swamp area of southern Delaware in 1929. It was rumored that the blazes were started by bootleggers operating in the area following a period of severe and prolonged drought. Until doused by heavy rains, the fires burned into the swamp's heavy peat layer that had built up over many years. In this photograph, a Mr. Fatzinger is standing beside a log that was covered with soil before the fire, clearly indicating the soil level before the fire took place. (Delaware Forest Service.)

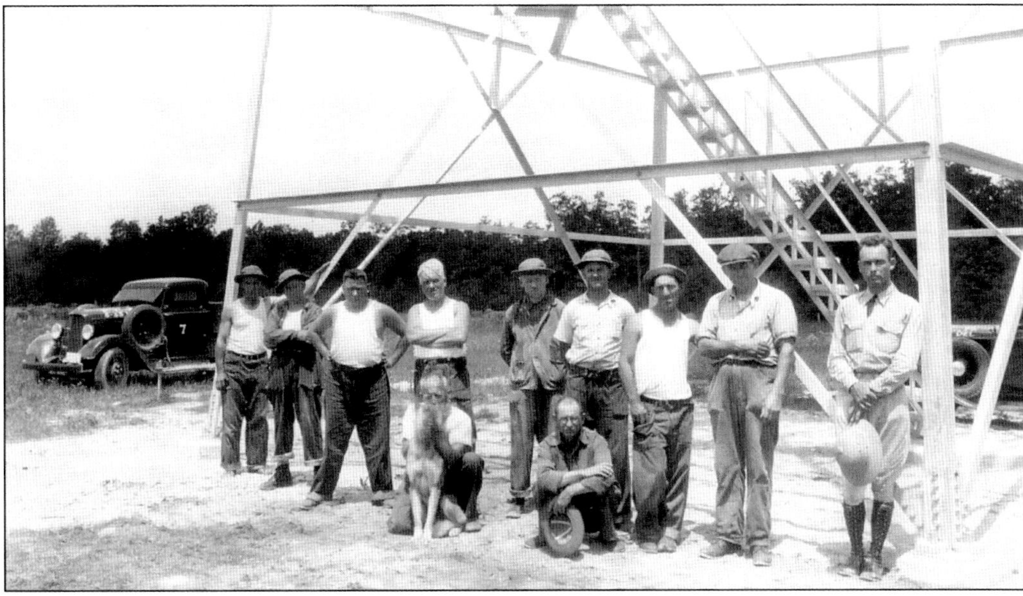

This 1935 photograph shows the men from Civilian Conservation Corps Camp S-53; they built Delaware's Viola Fire Tower. As part of the New Deal, the CCC was created as a public works relief program that operated from 1933 to 1942 for unemployed, unmarried men. In addition to building the system of fire towers, the crews also improved roads, built infrastructure, and undertook many reforestation and conservation projects. (Delaware Forest Service.)

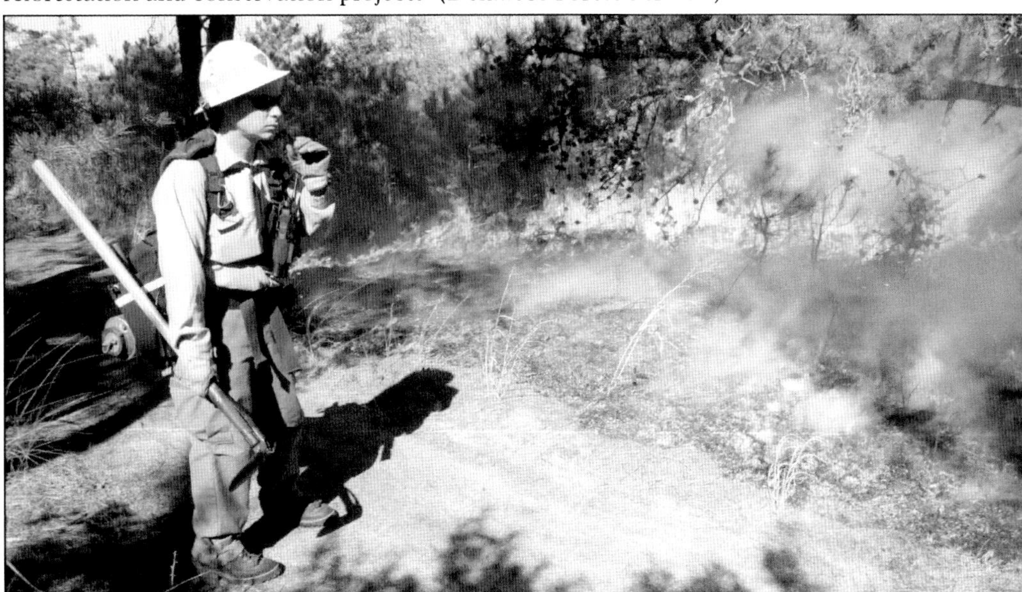

Samual Topper patrols the fire line of a controlled burn at Cape Henlopen State Park in 2012, as part of a wildlife and habitat enhancement project with the Delaware Division of State Parks. The plan was to reduce wildfire risk by burning fine fuels such as pine needles and dead branches while leaving larger woody debris unburned. This helped achieve the goal of reducing the overgrown shrub layer, opening the canopy to support rare shade-intolerant herbs and grasses, and consuming dead phragmites to enhance foraging habitat for young piping plovers. (Delaware Forest Service.)

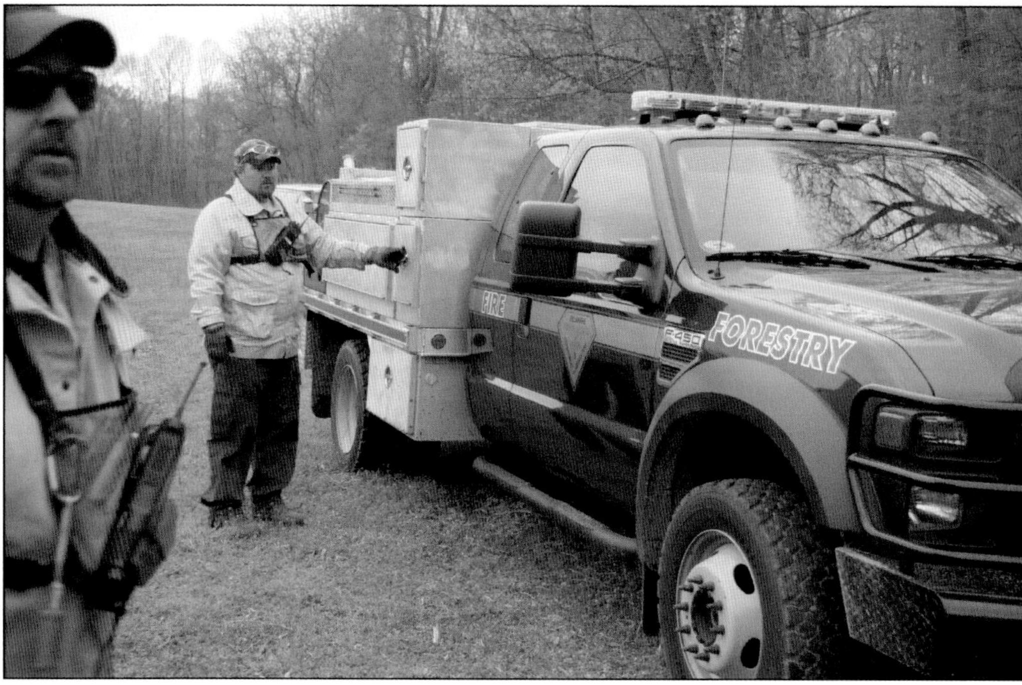

The Delaware Forest Service maintains two Type 6 engines—with one at Blackbird State Forest in the north and the other at Redden State Forest in the south. They are used for controlled burns, wildfire suppression, and to assist local volunteer fire companies when requested. This photograph was taken at the agency's annual "fire camp" training session held each spring, designed to train new Type 2 firefighters as part of national certification. Delaware regularly sends Type 2-IA wildfire crews on out-of-state assignments to support the national effort. (Delaware Forest Service.)

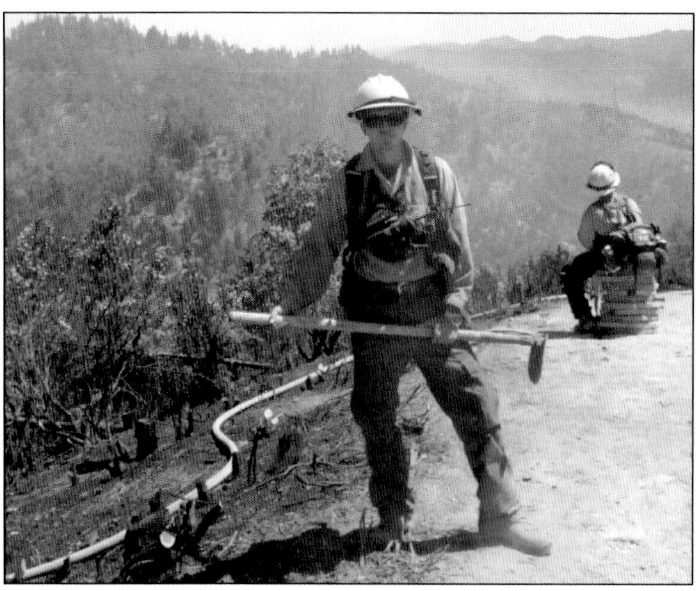

Kyle Hoyd is the Delaware Forest Service's assistant forestry administrator and oversees the agency's wildland fire program. This 2015 photograph shows him on the Delaware Type 2-IA crew that battled the Peak fire, one of many blazes caused by lightning that were part of the West Fork complex, which burned more than 35,000 acres near Hayfork, California. Each year, Delaware fields crews to mobilize through the National Interagency Fire Center. (Delaware Forest Service.)

Four

Great Lakes States

The Great Lakes states are comprised of Michigan, Minnesota, and Wisconsin. In 1983, fire managers from those three states met at the Indian Head Resort in Wakefield, Michigan, to discuss mutual concerns and needs regarding wildland fire protection. From that initial meeting, the Great Lakes Forest Fire Compact was organized. The state fire managers also requested the support of the US Forest Service's Northeastern Area State & Private Forestry to strengthen their partnership in providing wildland fire protection across the Great Lakes region. It was also agreed that the province of Ontario should also be a member of this organization. In September 1985, the states of Michigan, Minnesota, and Wisconsin signed the first mutual agreement, and the Great Lakes Forest Fire Compact was officially established. At this point, Ontario was a participant, but the formal international approval to join the compact came in 1989. In 1998, the province of Manitoba requested and was approved to become a member of the compact.

Over the years, the Great Lakes Forest Fire Compact has focused on mutual aid between the states and provinces by establishing committees to aid in coordinating firefighting operations, training, and wildfire prevention. These committees report to an executive board comprised of forest fire supervisors from each of the states and provinces. Since the initial organization's development, additional committees have been formulated that include air operations, law enforcement, and radio communications. Each addresses specialized components of preparing for and fighting wildland fires. The Great Lakes Forest Fire Compact's main goal is to promote resource sharing among its five member states and provinces. This resource sharing includes personnel, equipment, aircraft, information, and technology. All standing members gather annually in September in one of the states or provinces for a conference. They use this opportunity to build partnerships that continue to help make the Great Lakes Forest Fire Compact an international success.

The states and provinces within the Great Lakes Forest Fire Compact share common boundaries and topographic features. They have hazardous forest fuels that create unique forest fire protection challenges that warrant strong state and provincial fire protection agencies and programs. Historically, this region has been impacted by large wildfire conflagrations. Their fire protection programs have developed to be prepared for their states' or provinces' specific needs, along with assisting each other through suppression resource sharing during difficult times.

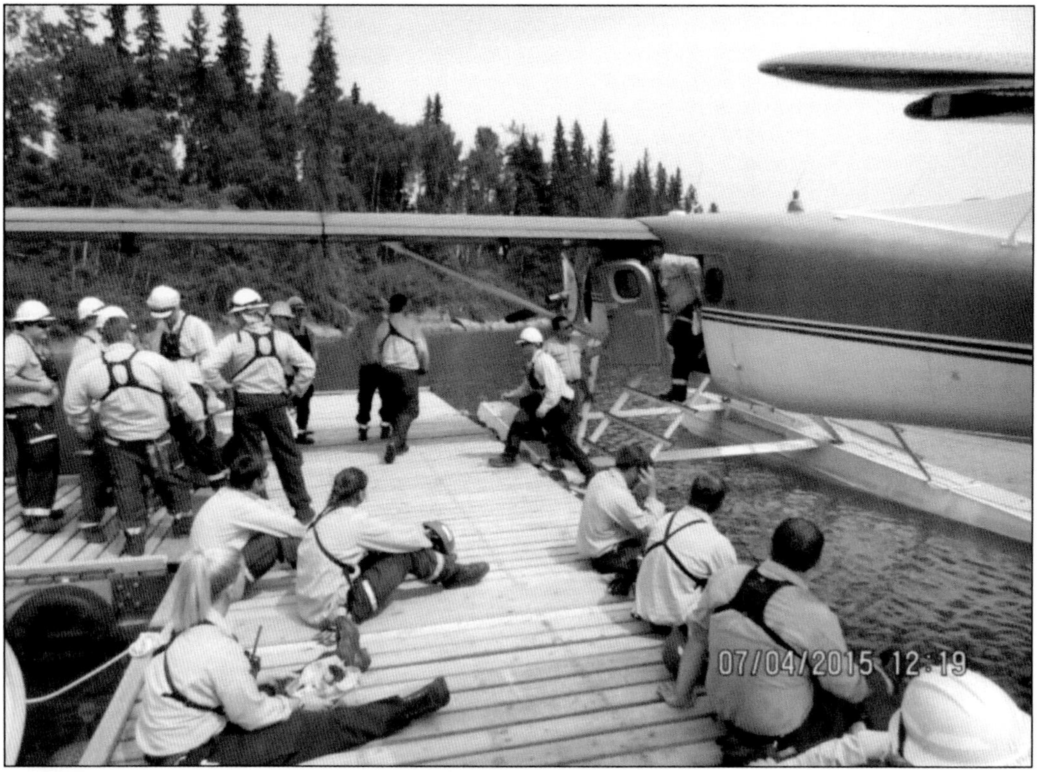

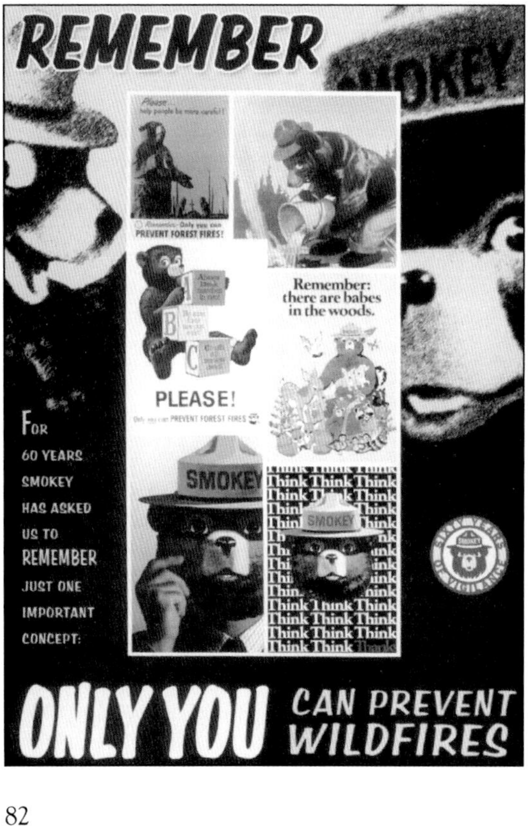

One of the major benefits of compact resource exchanges is learning different agency suppression tactics and bringing that knowledge back home. In 2015, Wisconsin sent a crew to Manitoba to respond to various fires across the northern portion of the province. Unlike Wisconsin, Manitoba has vast remote areas with abundant water and therefore utilizes float planes. Wisconsin crews were introduced to float plane safety and operations training to better understand how to work around fixed and rotary wing aircraft. (GLFFC.)

In 2006, the National Cooperative Forest Fire Prevention program presented the Great Lakes Forest Fire Compact with a Silver Smokey Bear Award for their continuous and outstanding service to wildland fire prevention. The compact was selected to receive the award because of their collective effort to develop a fire prevention education/awareness campaign and materials for the member states and provinces to use in celebration of Smokey Bear's 60th Birthday. (GLFFC.)

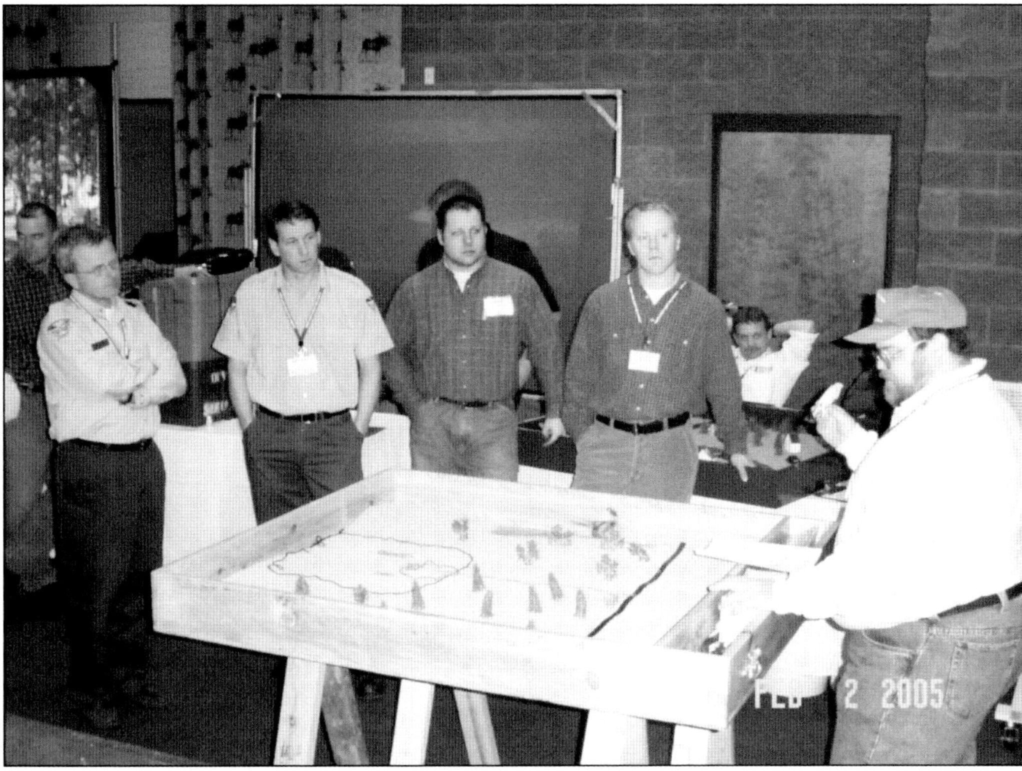

The Great Lakes Forest Fire Compact routinely holds training conferences around the region. These opportunities allow compact members to attend sessions sometimes not offered by the home agency. In 2005, Wisconsin hosted a Simulation Conference where students were given scenarios laid out on a tabletop sandbox. The role players then moved the equipment and described their plan of attack. Those managing the scenario incorporated more complexity, if needed, to guide the role players to the objective for the given scenario. (GLFFC.)

In 2008, the Great Lakes Forest Fire Compact recognized the need to adopt a stand-alone Law Enforcement Committee. One of their recent accomplishments was conducting a Railroad and Fire Investigation Training class in 2014, held in Duluth, Minnesota. The main goal was to build confidence when dealing with railroad fire issues. This included how to stop a train for inspection, how to safely work around the rails and locomotives, and the ins and outs of all the proper documentation. (GLFFC.)

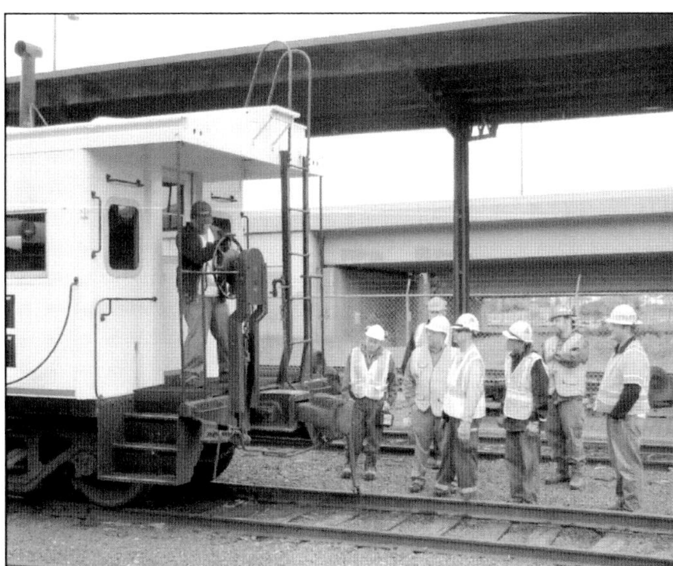

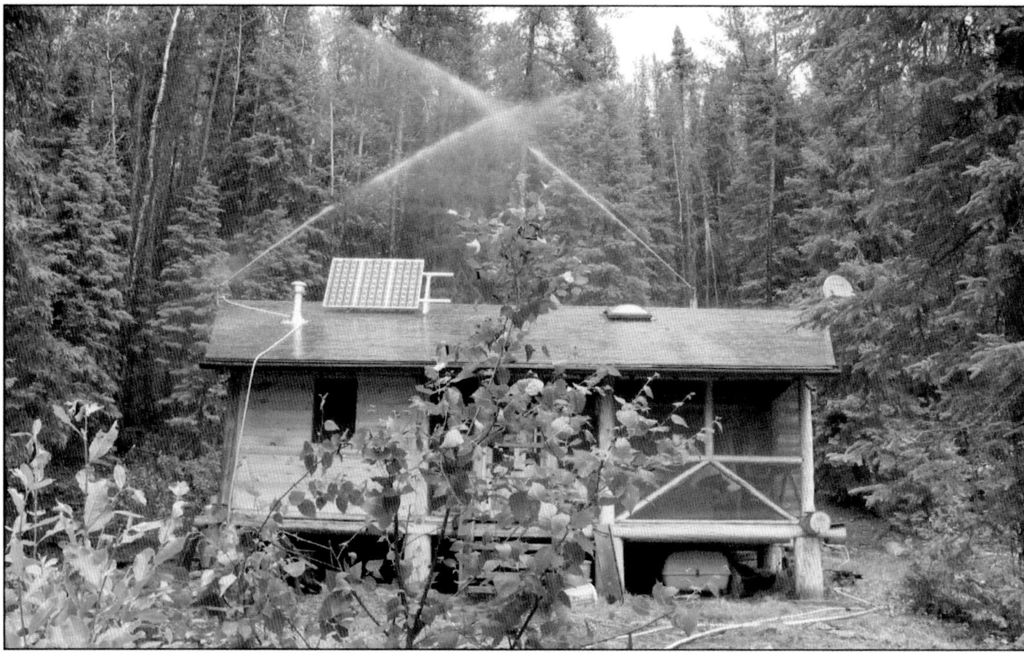

In 2003, the Ontario Mobile Values Protection Unit (consisting of a four-person crew and two technical experts) demonstrated their expertise and experience in structural values protection. The tour and hands-on demonstration proved to be quite informative and gave Wisconsin fire suppression staff a good understanding of how Ontario addresses wildfires threatening their wildland-urban interface. This photograph shows a structure protection sprinkler system in place and operating. (GLFFC.)

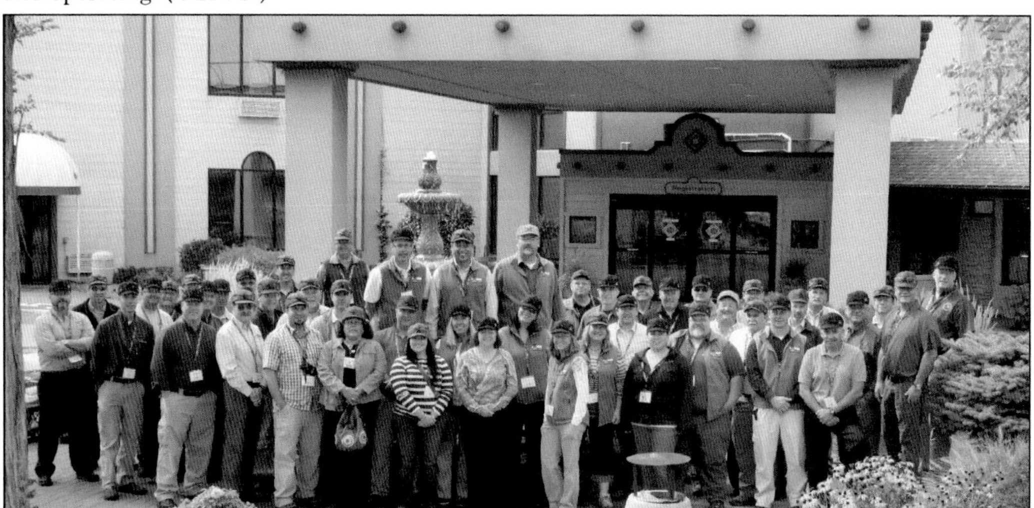

Every year, the Great Lakes Forest Fire Compact meets in September for the annual conference. The meeting location rotates among agencies, and the theme is established the prior year as identified by the executive board. In 2013, Wisconsin hosted the annual meeting in Wisconsin Dells, and the theme was "New Challenges Lead to New Opportunities." Some of the highlights included presentations on lessons learned from Texas drought impacts on live fuel moistures and fire behavior and an Aldo Leopold Foundation field tour. (GLFFC.)

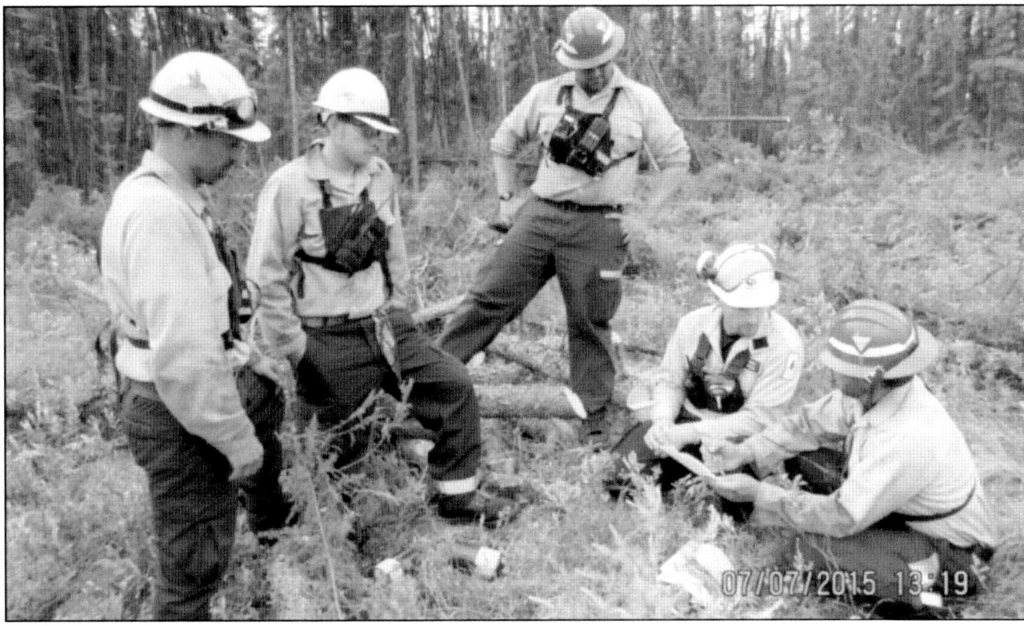

Mutual aid requests within the GLFFC allow for resource transfers to occur quickly, as the need arises. This was one of six Wisconsin crews working around Pikwitouie Lake on a 12,000-acre fire in Manitoba in July 2015. Four crews were spiked out and forced to be self-sufficient in organizing and maintaining their campsites with supplies brought in daily by helicopter. Two other crews worked initial attack by helicopter from two remote ranger station sites. All crews fought fire the "Canadian way," using large-scale hose lays. (GLFFC.)

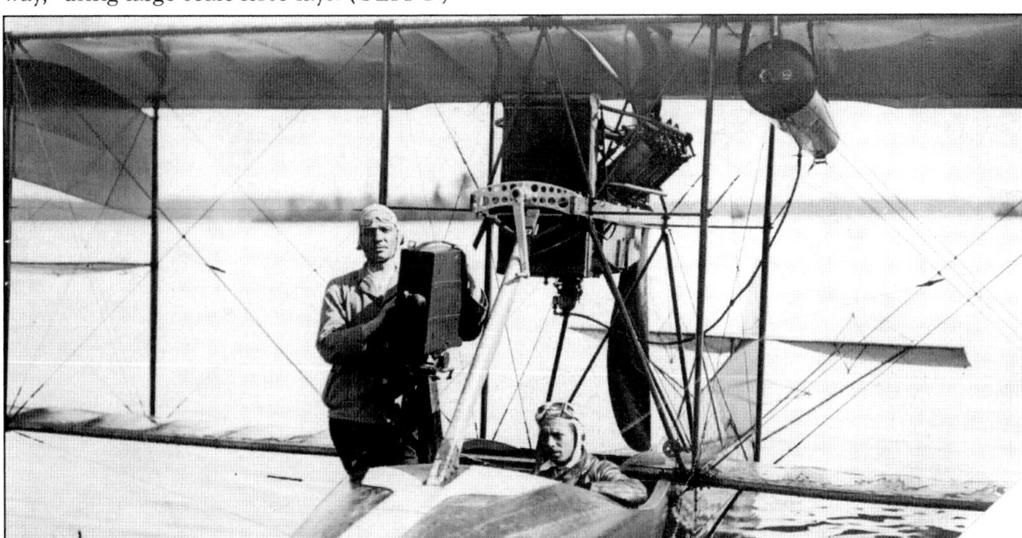

On June 22, 1915, aviation pioneer Jack Vilas, with the assistance of state forester E.M. Griffith, got the idea to use his Curtiss flying boat to assist Trout Lake rangers at the Wisconsin Conservation Department spot fires from the air and catch them before they got too large. Upon landing, the crew immediately reported the smoke's direction and location to the rangers. Flight detection was catching on across the continent, and by 1917, the US Forest Service implemented what they called the "Wisconsin plan" for fire detection. (Wisconsin DNR Division of Forestry.)

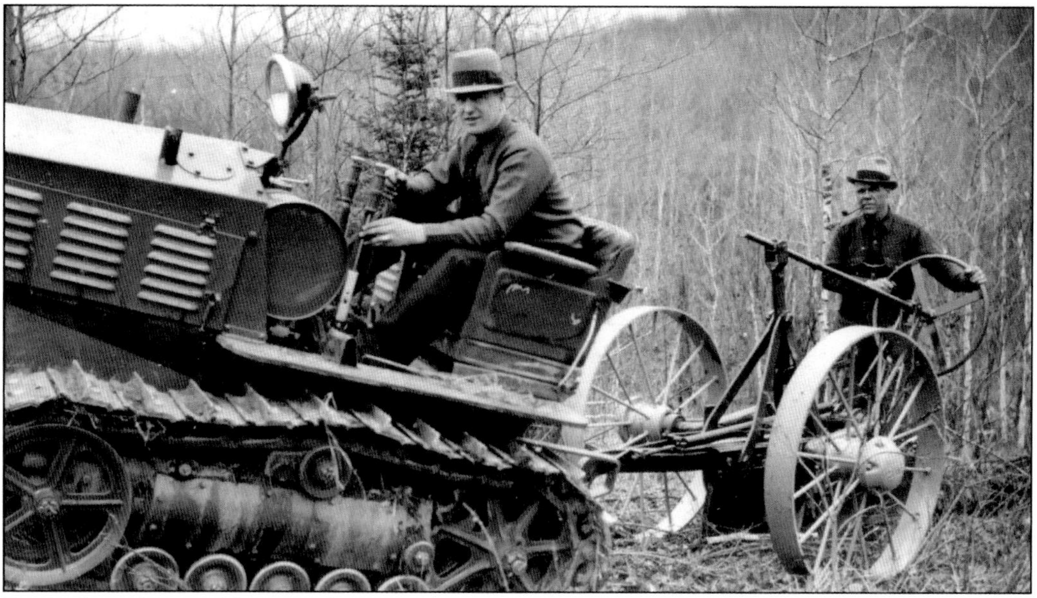

By the late 1920s, public attitudes toward fire shifted. What once was tolerated became a menace and a threat to public safety and Wisconsin's northern economy. In 1927, the legislature approved a plan for a complete overhaul of the state's fire protection system. The Wisconsin Conservation Department assumed full responsibility for fire protection on all land in fire districts. Rangers introduced innovative firefighting techniques and built some of the equipment themselves, such as this Cletrac with case plow and one of the first tiltbed trailers. (Wisconsin DNR Division of Forestry.)

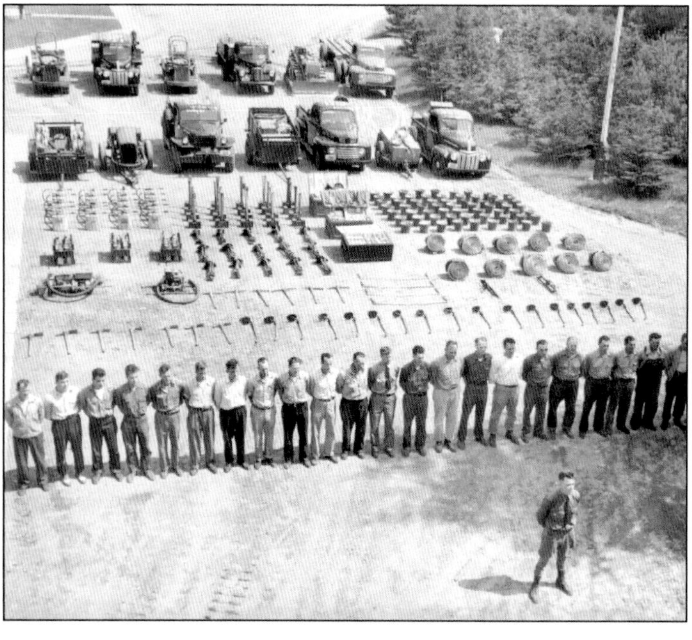

Early fire wardens, rangers, and the Civilian Conservation Corps (CCC) battled forest blazes with simple hand tools. The success of their endeavors often depended on how many hands were available to wield those tools. However, after the CCC disbanded in 1942, finding enough people to fight the flames became a problem. The answer was to organize a smaller but better trained group of volunteer firefighters and equip them with the latest fire control machinery, as seen here at the Mercer Ranger Station in 1949. (Wisconsin DNR Division of Forestry.)

Between 1905 and 1911, forest fire control was marked by appointing 249 town fire wardens around the state. In addition, 12 forest rangers were trained, certified, and hired, and the forest protection headquarters was established at Trout Lake. From this point through the late 1920s, organized protection spread across the state as ranger stations and fire lookout towers were constructed. (Wisconsin DNR Division of Forestry.)

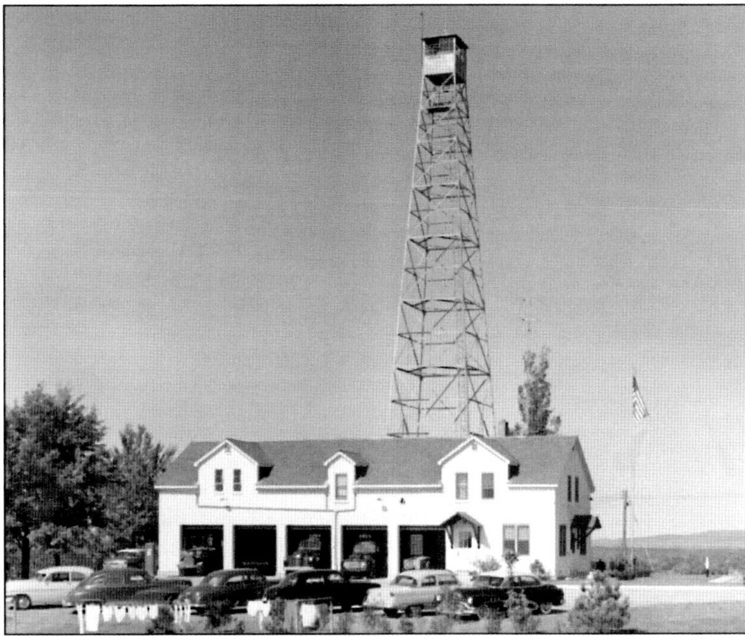

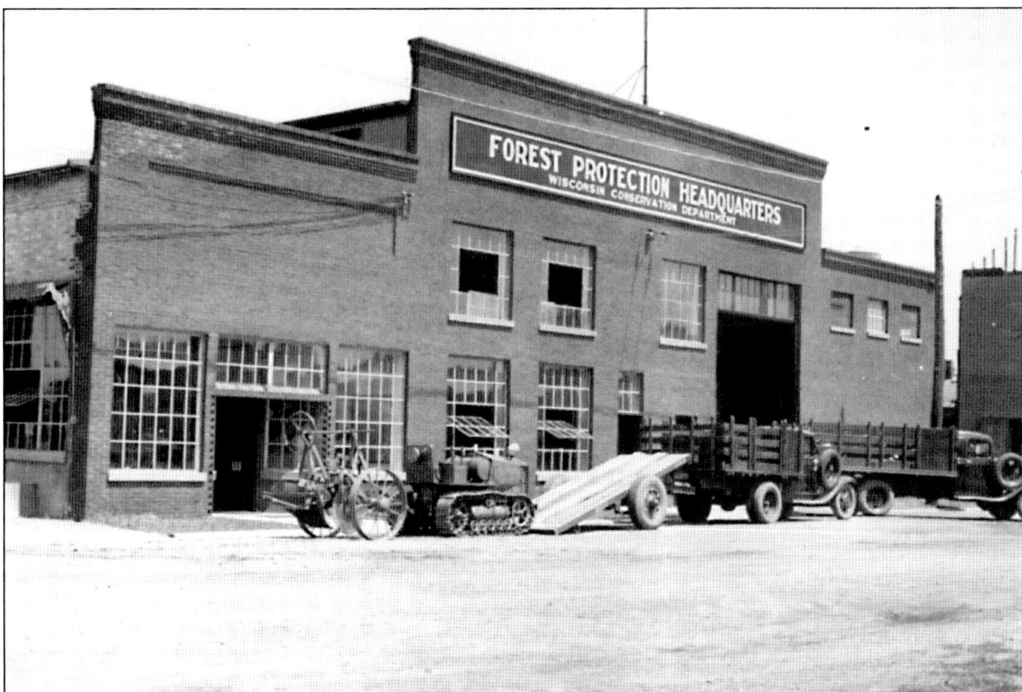

In 1934, Wisconsin established a central warehouse and construction and repair shop, as well as a forest equipment testing and experiment station. Centered in Tomahawk, Wisconsin, the LeMay Forestry Center (formerly known as the forest protection headquarters) continues to be the headquarters for doing the mechanical, lathe, blacksmith, welding, carpentry, and paint work necessary for Wisconsin's fire program. All heavy and hand equipment used in fire suppression are received and distributed from this historic location. (Wisconsin DNR Division of Forestry.)

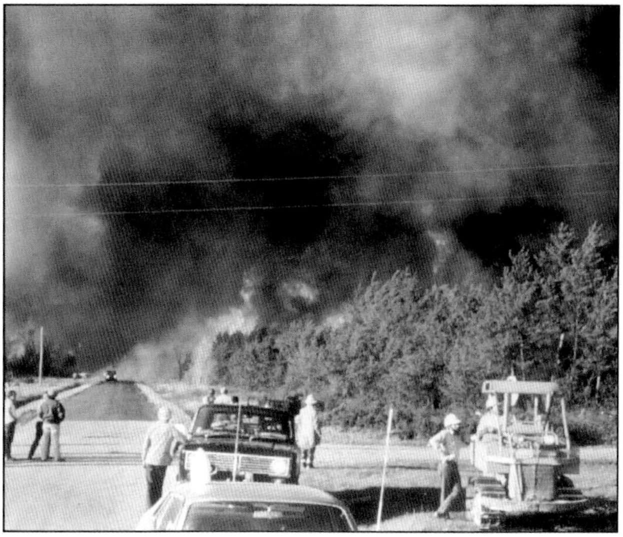

In 1977, Wisconsin suffered a second year of severe drought, and nearly 50,000 acres burned. Over 170 structures were destroyed or damaged. The notable Saratoga fire burned 6,159 acres, and nearly 100 structures were lost. On the same day, a series of fires caused by trains combined to form one large fire, named the Brockway fire. Later that day, an illegal cooking fire escaped and joined the railroad fires. The Brockway fire was finally controlled after burning 17,590 acres with damages over $1.4 million, including the loss of 14 homes. (Wisconsin DNR Division of Forestry.)

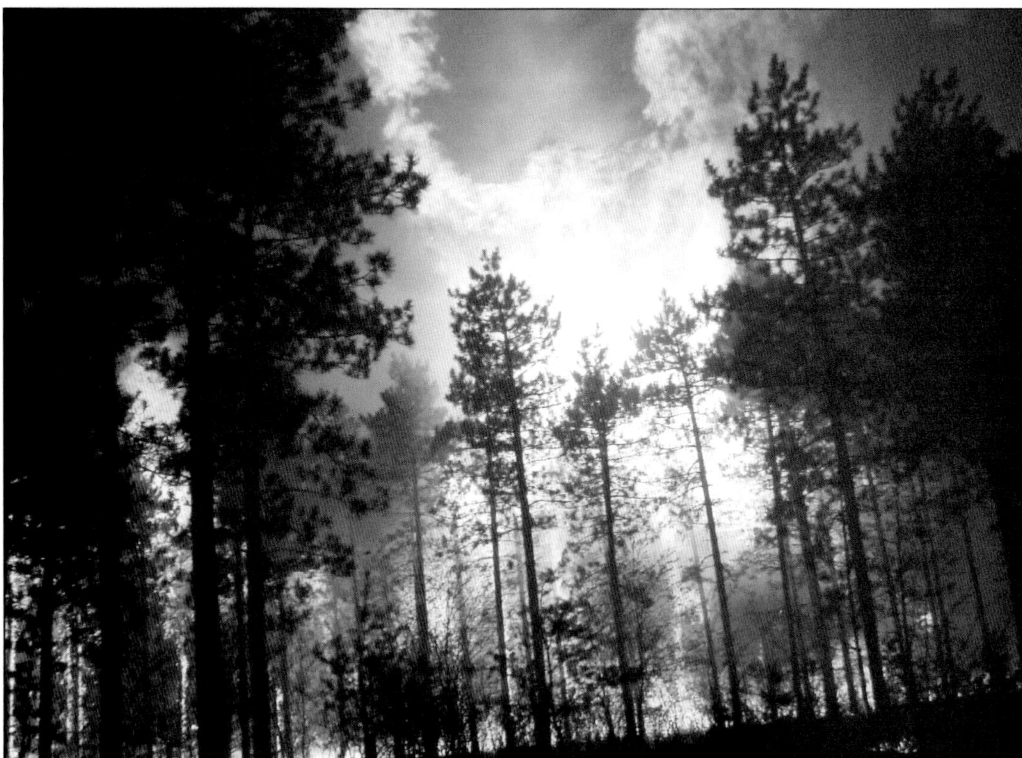

The Cottonville fire began on May 5, 2005, in northern Adams County and rapidly spread. Resources included 38 tractor plows, 25 forest rangers with Type 7 four-by-fours, three low-ground units, six heavy dozers, and almost 200 DNR personnel to control and suppress the wildfire. Air resources cooled the flanks for ground crews and dropped water and fire retardant on structures. The fire was finally contained after 3,410 acres had burned. Thirty homes and at least 60 outbuildings were destroyed, but an estimated 300 buildings were saved. (Wisconsin DNR Division of Forestry.)3

The Germann Road fire in northwest Wisconsin occurred on May 14, 2013, burning an average of 1,000 acres per hour. By 8:00 a.m. on May 15, crews had 94 percent (33 miles) of the fire contained; this included nearly 24 miles of dozer-constructed fire line, 6 miles of natural barrier, and 3 miles of direct suppression with hoses. In the end, 7,499 acres burned and over 104 structures were lost. This was the largest fire in Wisconsin in over 30 years. (Wisconsin DNR Division of Forestry.)

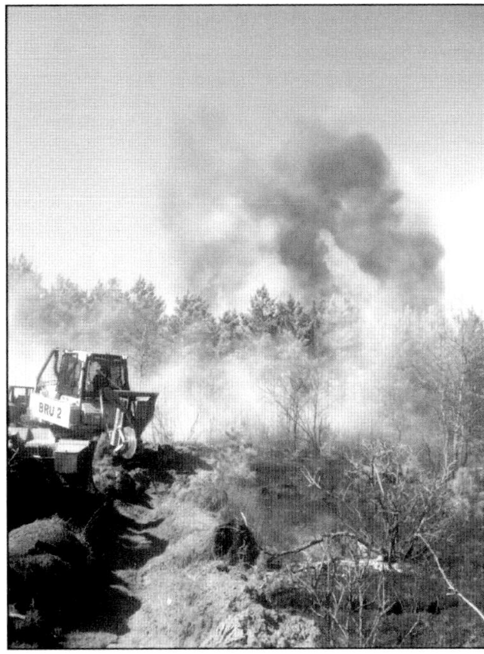

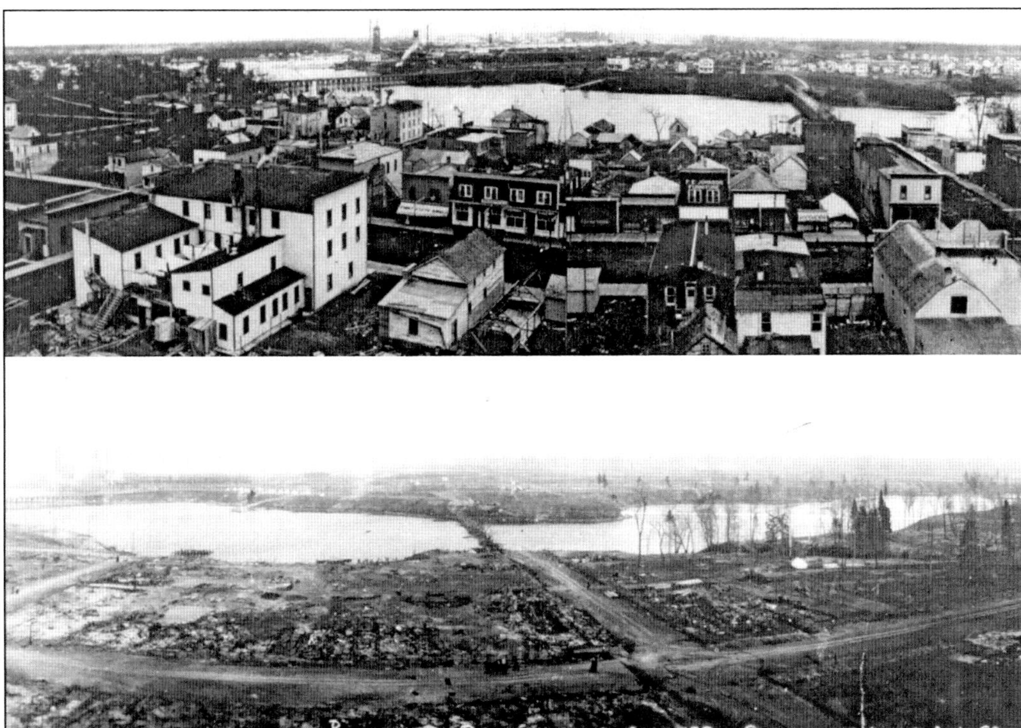

The Baudette-Spooner fire of 1910, which incinerated six cities, consumed 300,000 acres, and killed 42 people, precipitated the creation of the Minnesota Forest Service in 1911. Twenty newly hired forest rangers were charged with all the fire control work of the federal government, lumber companies, railroads, and private citizens in the forested parts of Minnesota. These photographs depict before and after photographs of Baudette-Spooner. (Minnesota DNR Division of Forestry.)

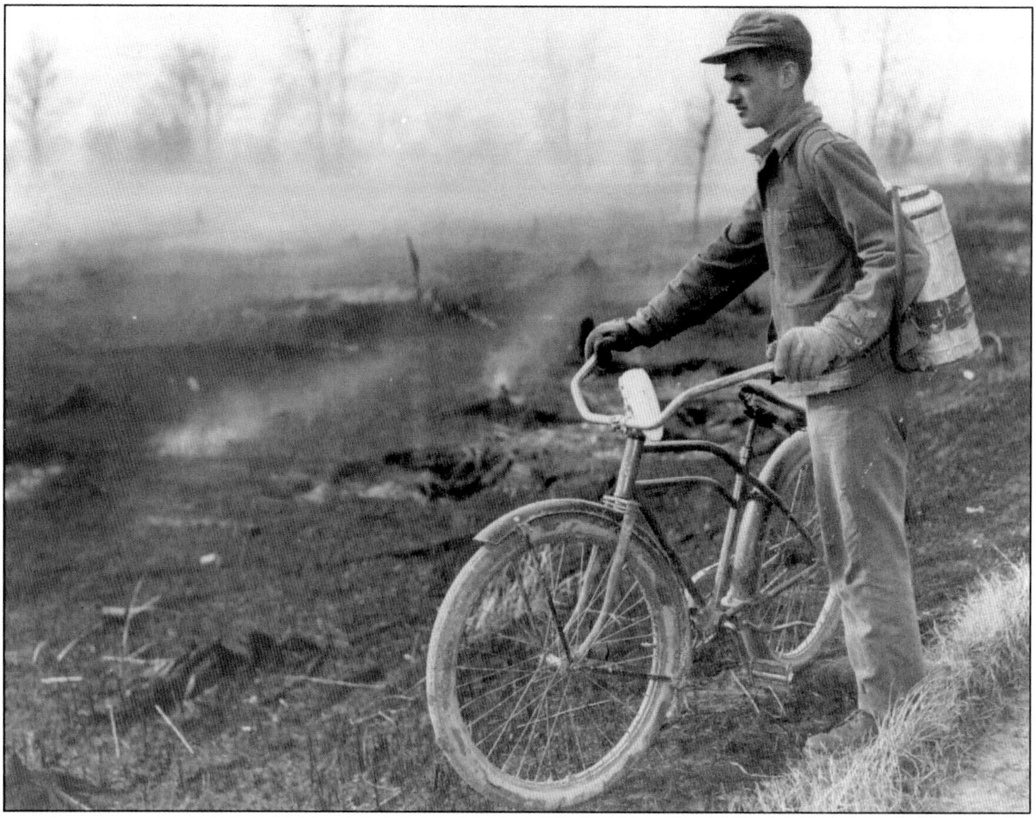

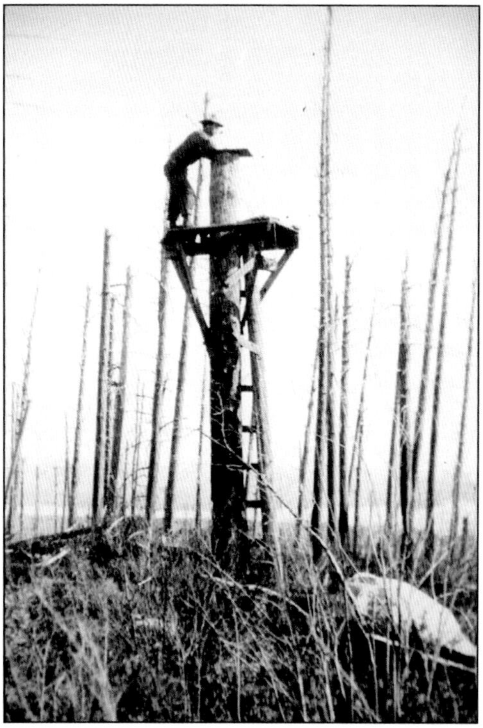

In the earliest days, many modes of transportation were used for fire patrol and response throughout Minnesota. Horses and drays were often used on peat fires. Motorcycles, with strapped-down fire tools, provided quick response times and the ability to drive over terrain inaccessible to larger vehicles. As shown in this photograph, bicycles were frequently used to map fire perimeters. (Minnesota DNR Division of Forestry.)

Minnesota fire towers improved structurally as time went forward. The first towers were often simple wooden platforms elevated above its surroundings, illustrated here by the Birch Lake Fire Tower constructed around 1918. John Fritzen, a North Shore ranger, stated, "The (Minnesota) Forest Service had no equipment, not even hand tools. The detection system, consisting of a few wooden scaffolds, was almost nonexistent." (Minnesota DNR Division of Forestry.)

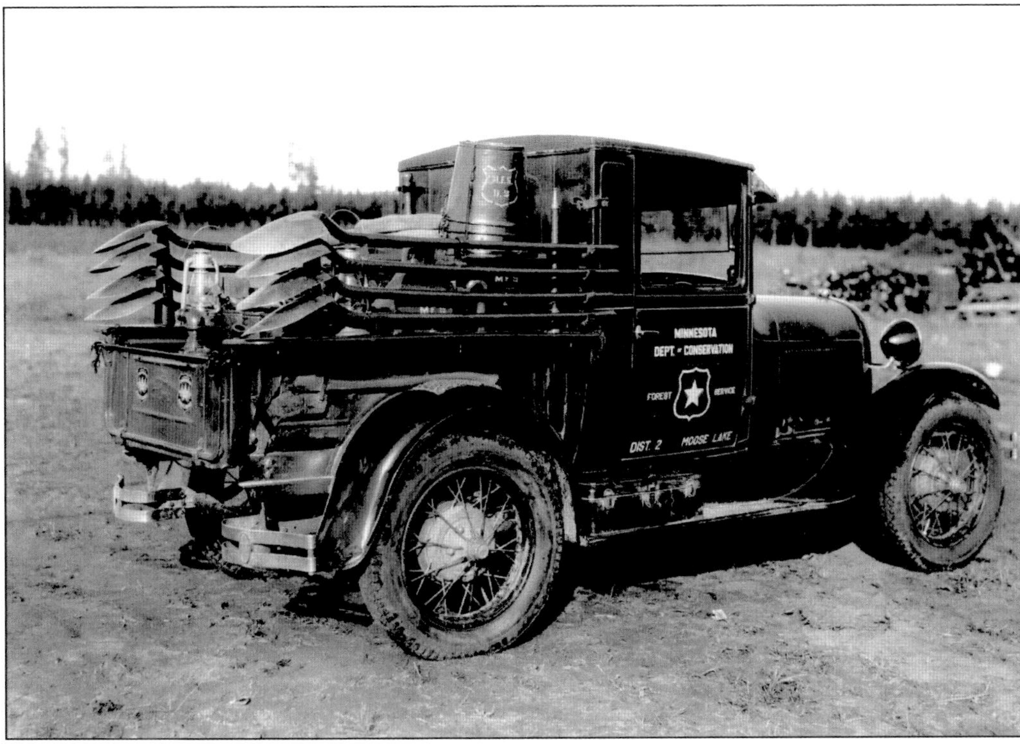

This 1928 Model A fire truck was fully equipped and ready to respond to fires in the 1930s. In those days, the most common method of fighting non-peat fires was with the aid of a shovel and, when possible, buckets of water. By the next decade, the Minnesota Department of Conservation had upgraded to slip-on pumps and hoses that were put into three-quarter-ton pickups. (Minnesota DNR Division of Forestry.)

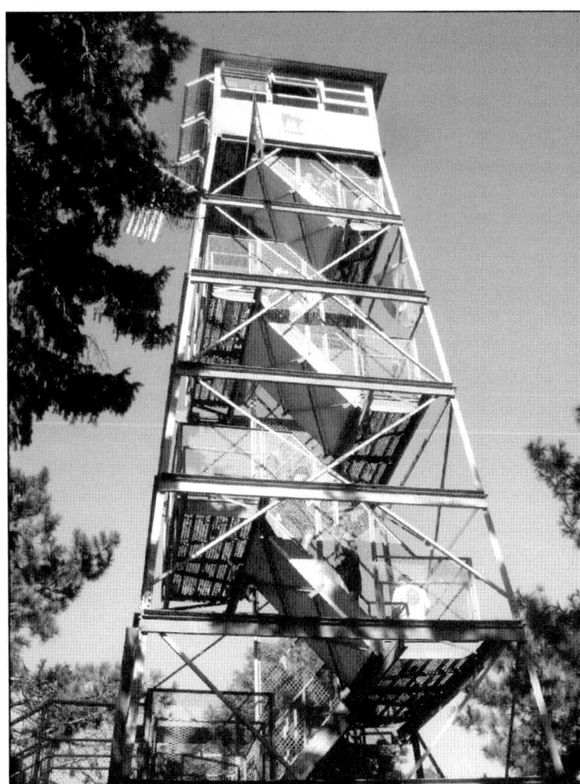

Though infrequently, fire towers are still used today for smoke detection. The most popular tower is a 65-footer located in St. Paul on the Minnesota State Fairgrounds. This tower is only used by the Minnesota DNR for educational purposes but is open for the entire run of the 12 days of the state fair. About 1,000 fairgoers climb the tower daily. (Minnesota DNR Division of Forestry.)

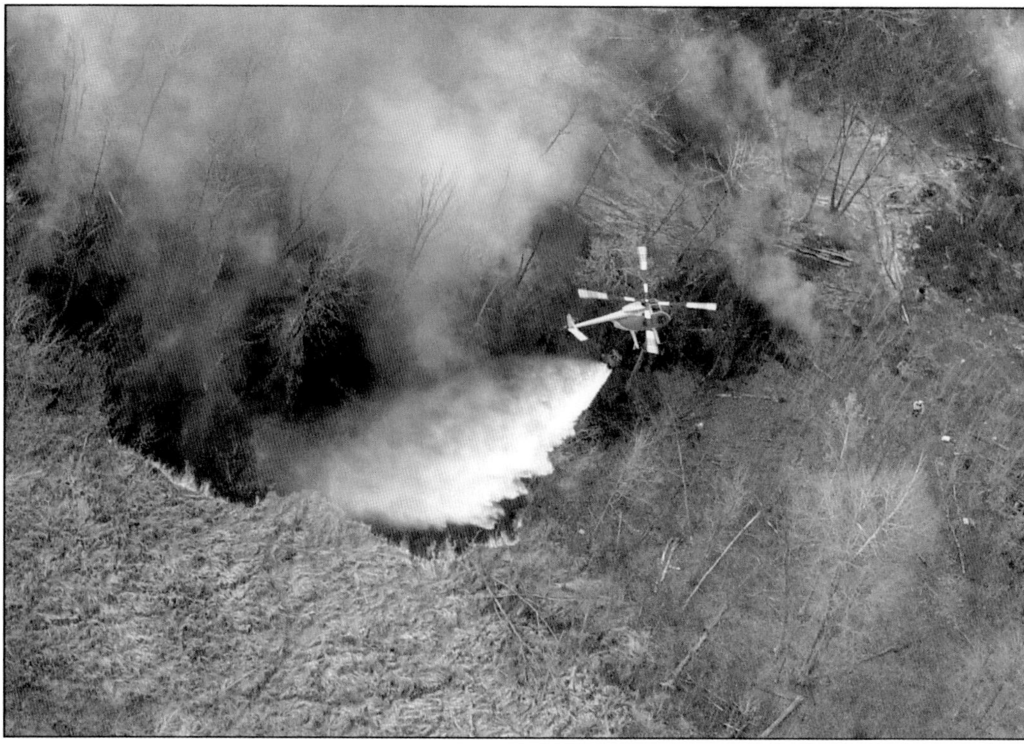

A severe drought in 1976 forced Minnesota to change the way it fought wildfires—with aviation resources, helicopters, and air tankers. Through innovation, determination, and dedication, the state created a top-notch air operations program. Helicopters still play a huge role in wildfire suppression and forest management projects, including aerial seeding of spruce and pine species on cutover areas and slinging loads of trees to remote planting sites. (Minnesota DNR Division of Forestry.)

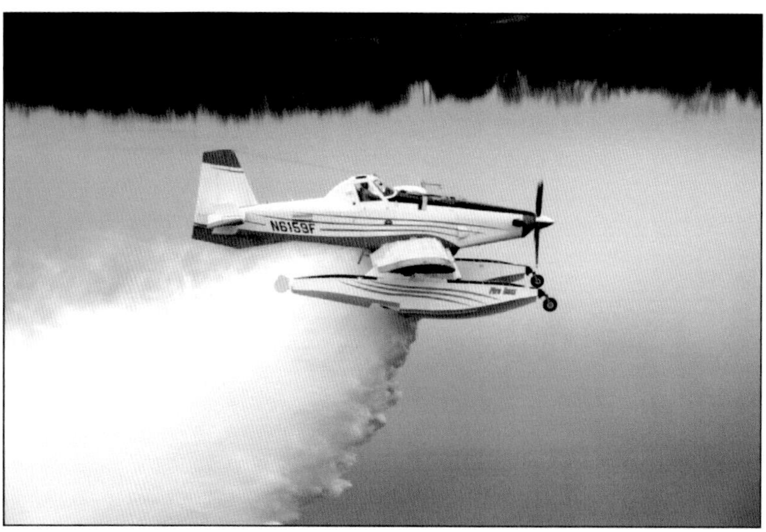

Showing their age, in 2014 the CL-215s were sold by Minnesota DNR and replaced with contracted ground-based single-engine air tankers (SEAT) and a new version of water-scooping aircraft, the Fire Boss. These versatile planes have nearly an 800-gallon capacity and can deliver either water or retardant on a fire. (Paul Woods.)

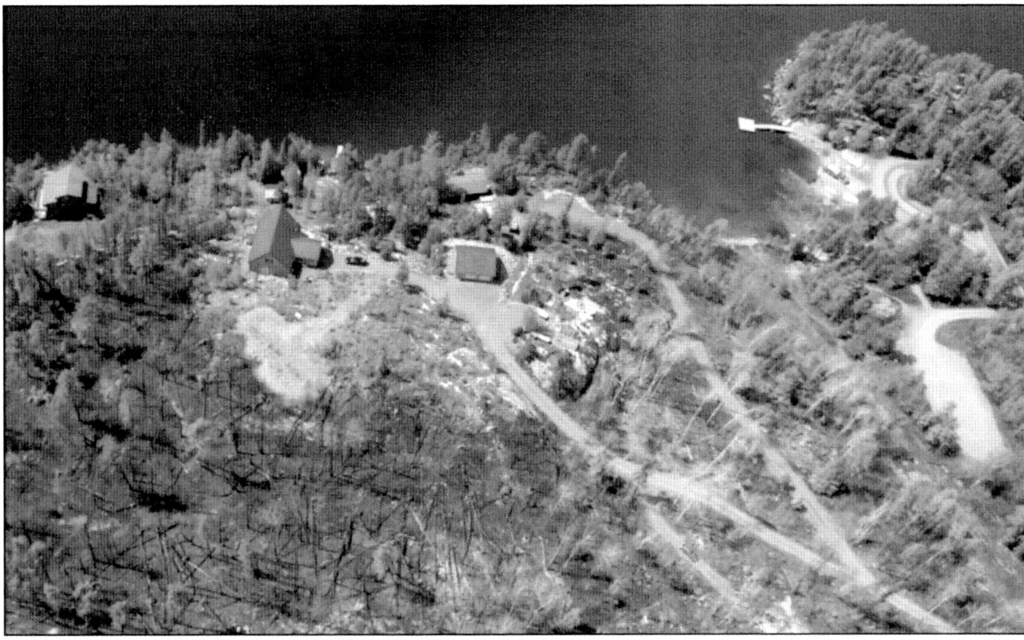

The Minnesota DNR has been active in the national Firewise program since 2001. There has been a lot of success with the program, but it often intermingles with the "it won't happen here" attitude. Unfortunately, it sometimes takes a catastrophic fire before a community takes action. This aerial photograph shows a mix of homes that survived the Ham Lake fire and others that did not. The 2007 Ham Lake fire burned 75,443 acres across Minnesota and Ontario. One hundred thirty-eight homes were destroyed by the fire in Minnesota and 15 in Ontario. (Minnesota DNR Division of Forestry.)

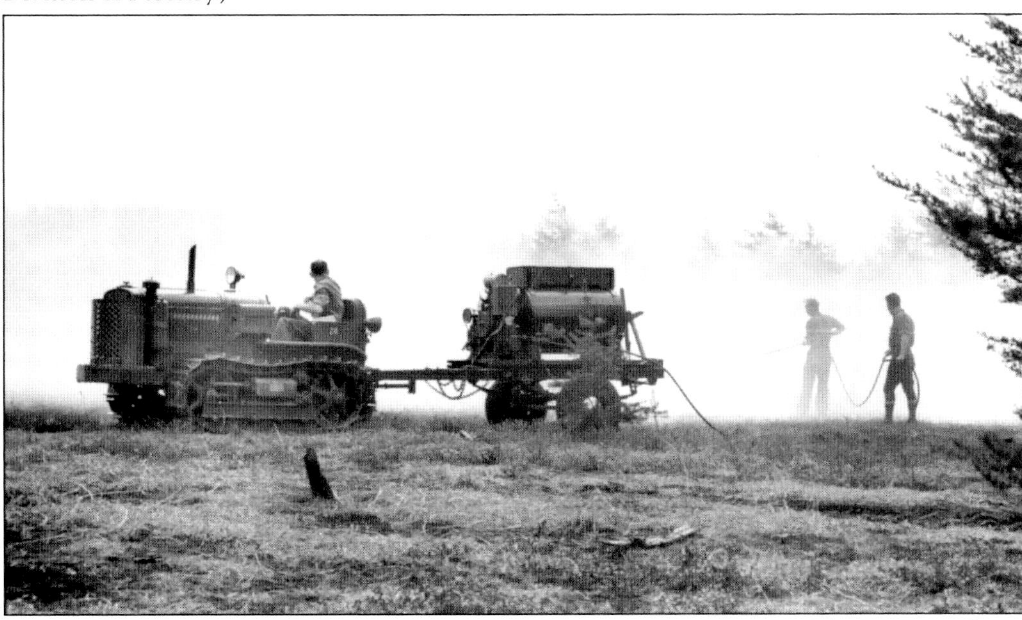

Pictured is a 1940s vintage dozer with water tank and pump trailer. Firefighting in Michigan has evolved through many stages of tanks, pumps, dozers, tractor plows, engines, and skidders (skidgines). (Michigan DNR Forest Resources Division.)

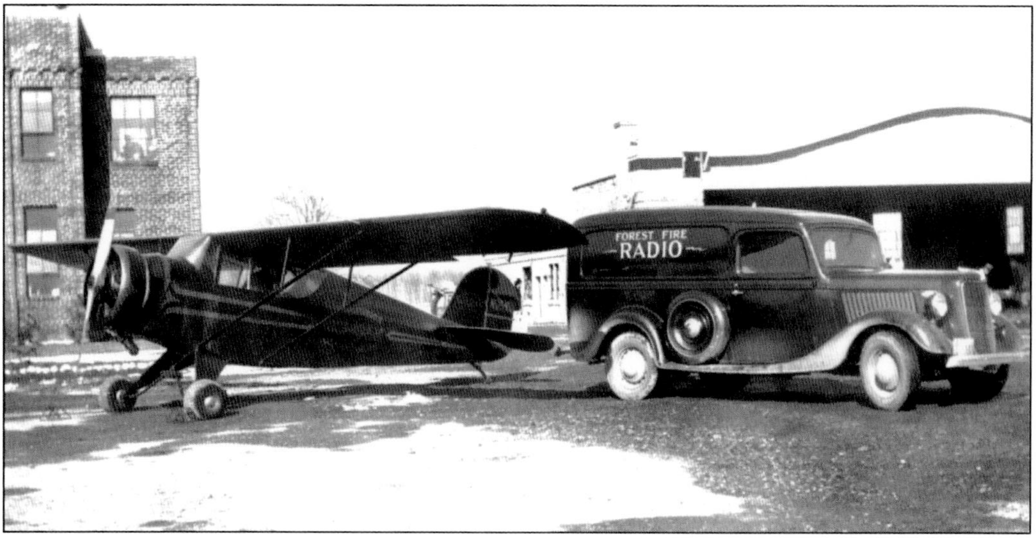

This 1935 photograph shows a Michigan DNR detection aircraft and radio truck. This is one of the earliest such combinations of specialized resources for reporting fires and dispatching firefighters in Michigan. (Michigan DNR Forest Resources Division.)

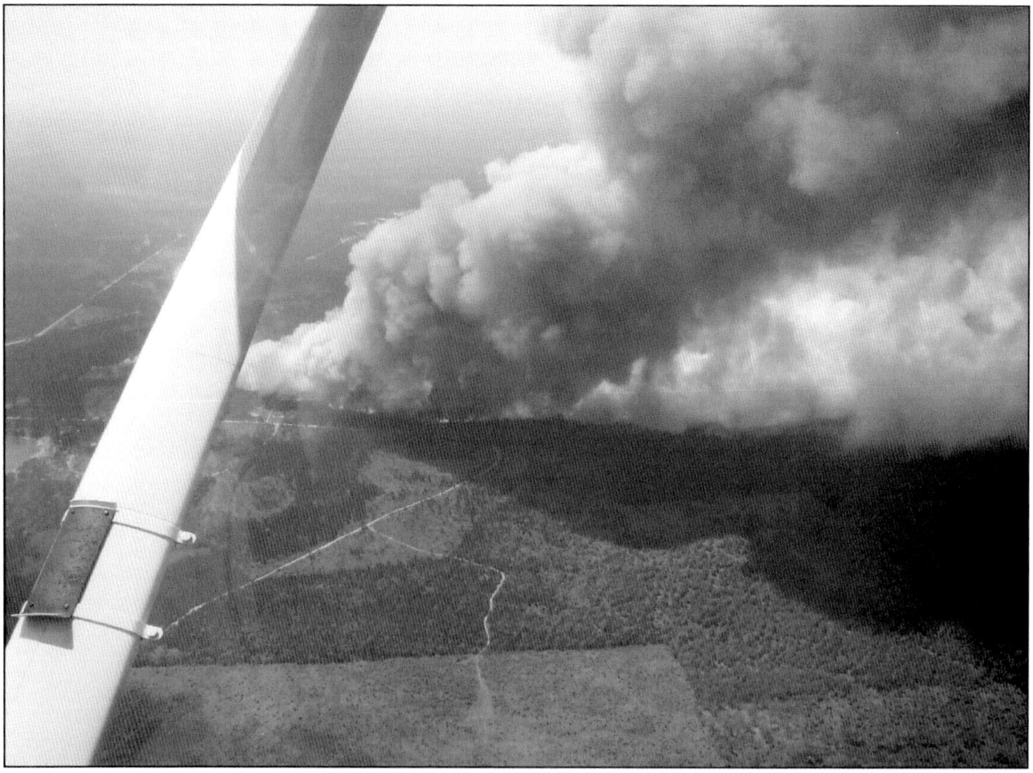

This aerial photograph shows a Michigan wildfire burning in mature and open Jack pine. This forest type is considered one of the most volatile fuel types in the Midwest. Typically, the Jack pine trees candle or develop new growth in the spring, resulting in a significant drop in live fuel moisture and high rates of fire spread. Michigan utilizes aircraft for both fire detection and fire intelligence duties. (Michigan DNR Forest Resources Division.)

The 4 Mile Road fire occurred April 24, 2008, and threatened the city of Grayling, Michigan. The fire burned over 1,300 acres and was caused by railroad activities. The fire jumped both the north and south lanes of Interstate 75, threatened over 85 structures, and burned 3 residences and 3 outbuildings. This view of the fire is from a parking lot on the business loop of Interstate 75 in Grayling. (Michigan DNR Forest Resources Division.)

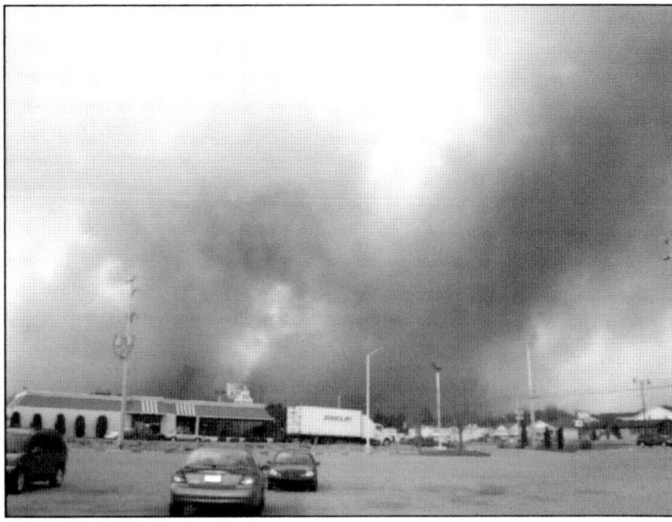

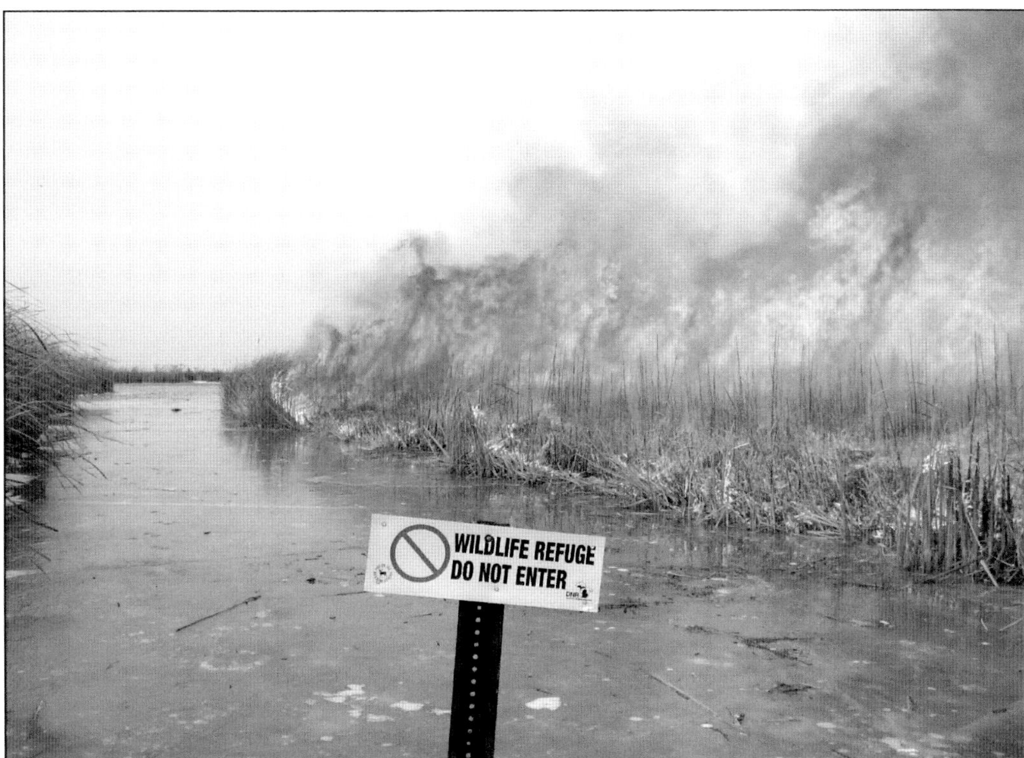

The Michigan DNR Forest Resources Division conducts approximately 100 prescribed fires a year for over 10,000 acres. Michigan DNR burns in a variety of fuel types for wildlife habitat improvements, such as these cattails; for invasive species management; for forest and natural communities' regeneration; and for fuel reduction. (Michigan DNR Forest Resources Division.)

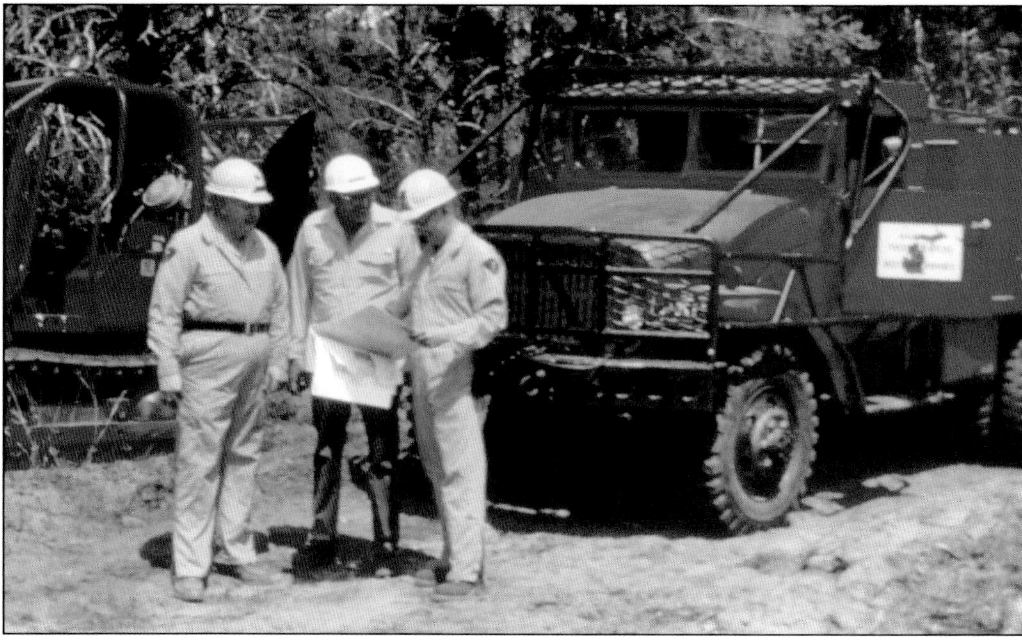

Pictured in the 1970s, Michigan DNR firefighters discuss firefighting tactics, with a tractor plow and military surplus six-by-six engine in the background. The Michigan DNR Forest Resources Division utilizes a variety of federal surplus equipment and develops them into fire response equipment at the Roscommon Forest Fire Experiment Station. (Michigan DNR Forest Resources Division.)

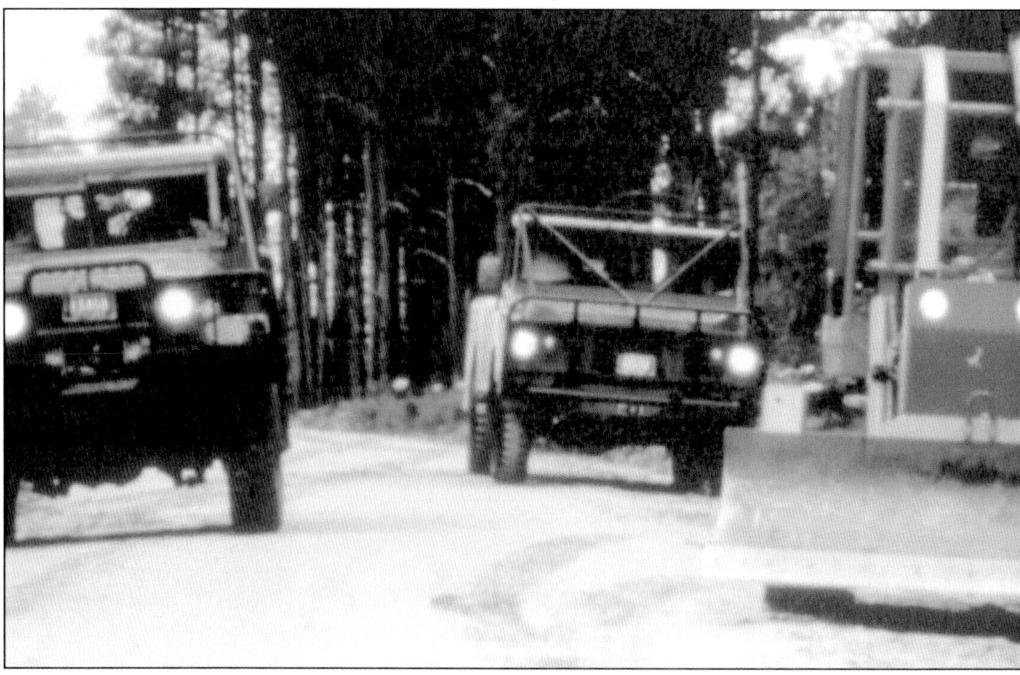

Staged on the road near a fire is a group of very typical Michigan firefighting equipment. From left to right are the military six-by-six engine with 1,500 gallons of water; the International four-by-four engine with 800 gallons of water; and the dozer with a Michigan fire plow mounted on the rear, which puts in a six-to-eight-foot plow line. (Michigan DNR Forest Resources Division.)

Five

Big Rivers States

The Big Rivers states are comprised of Illinois, Iowa, Missouri, and Indiana. The concept of forming a compact or other localized wildland fire organization within this region began in 1988 when the state foresters agreed to form the Midwest Alliance. The purpose of this group was to help promote wildland fire training and prevention among the member states in cooperation with the US Forest Service's Northeastern Area State & Private Forestry. The Midwest Alliance was recognized by the US Forest Service and was provided funding through the cooperative fire grant process. In 1991, the alliance received funds to develop a wildland-urban interface safety video.

In 1992, the Midwest Alliance felt that this trial of a multistate wildland fire joint organization was working quite well, but it lacked the permanency needed to be truly effective. In 1992, the state foresters from the Big Rivers states were requested to support an initiative to develop the alliance into a full-fledged fire compact. Each state had a different set of rules and regulations regarding the creation of a compact, and attempting to get a legislative mandate through each of the states would have been a time-consuming process. It was soon determined that the Weeks Act of 1911 provided the necessary framework for states to form forest fire compacts. In 1995, the Big Rivers Forest Fire Management Compact (BRFFMC) was officially formed.

The compact was created to protect human life, real property, and natural resources on lands under protective authority of the member states. The primary objectives were to promote and maintain effective fire management and common services and to develop an integrated all-risk response plan using mutual aid to assist member states. It was decided that much of the cooperative work would be through shared training and fire prevention efforts. In 2002, the Midwest Wildfire Training Academy was established. By the end of 2015, the academy had trained in excess of 4,500 students and provided over 250 courses.

The states within the Big Rivers region share the Mississippi and Ohio Rivers. Their natural vegetation ranges from hardwood forests in the southern areas to vast expanses of oak savannas and prairies along its northern and western boundaries. Fire is a natural part of these ecosystems. The states' fire management programs have evolved to manage wildland fires and use prescribed fire to meet public safety and land management objectives.

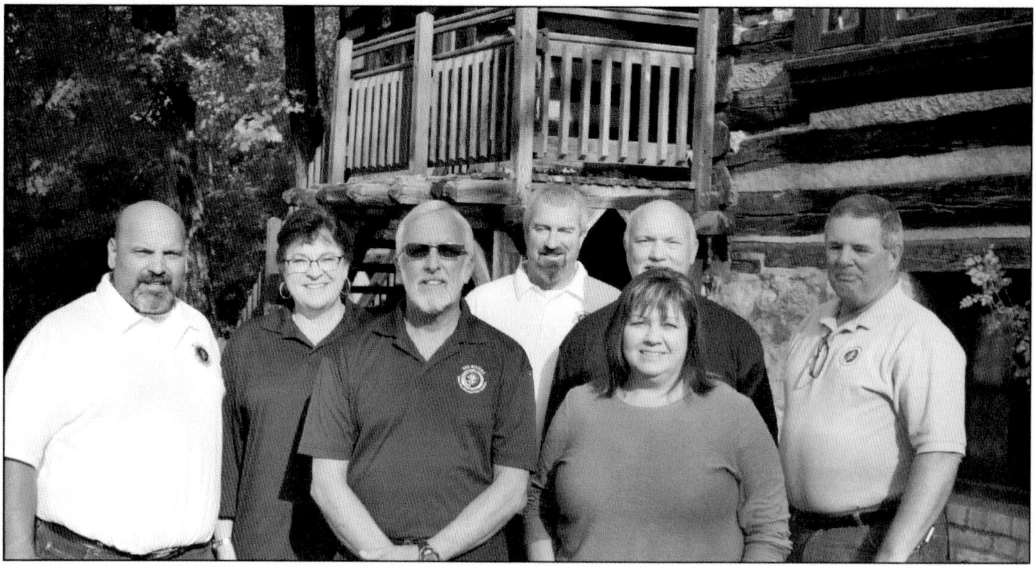

Members of the 2016 executive board of the Big Rivers Forest Fire Management Compact take a break and pose for a photograph at their fall meeting. This meeting was held at the Pere Marquette State Park in Illinois near the confluence of the Illinois and Mississippi Rivers. The cabin in the background was constructed by the Civilian Conservation Corps in the 1930s. (BRFFMC.)

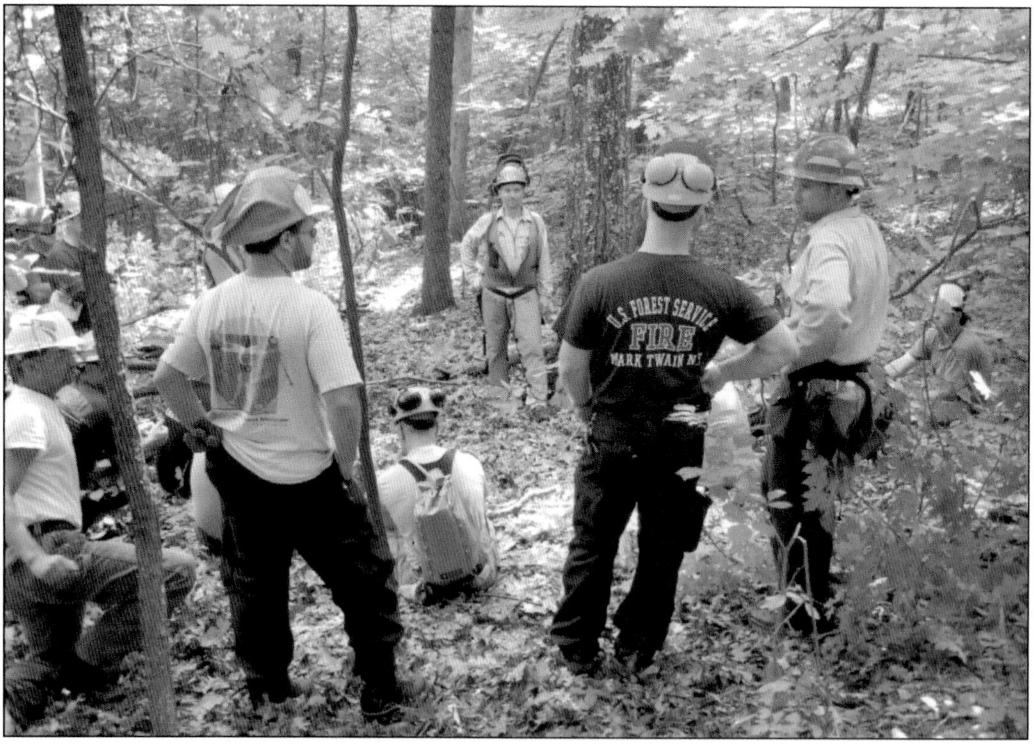

In 2002, the Big Rivers Forest Fire Management Compact hosted the first Midwest Wildfire Training Academy to meet the needs of wildland firefighters across the compact area. Through 2015, the academy has trained over 4,500 students in more than 250 courses. This photograph captures students and instructors during a wildfire chain saw training class. (BRFFMC.)

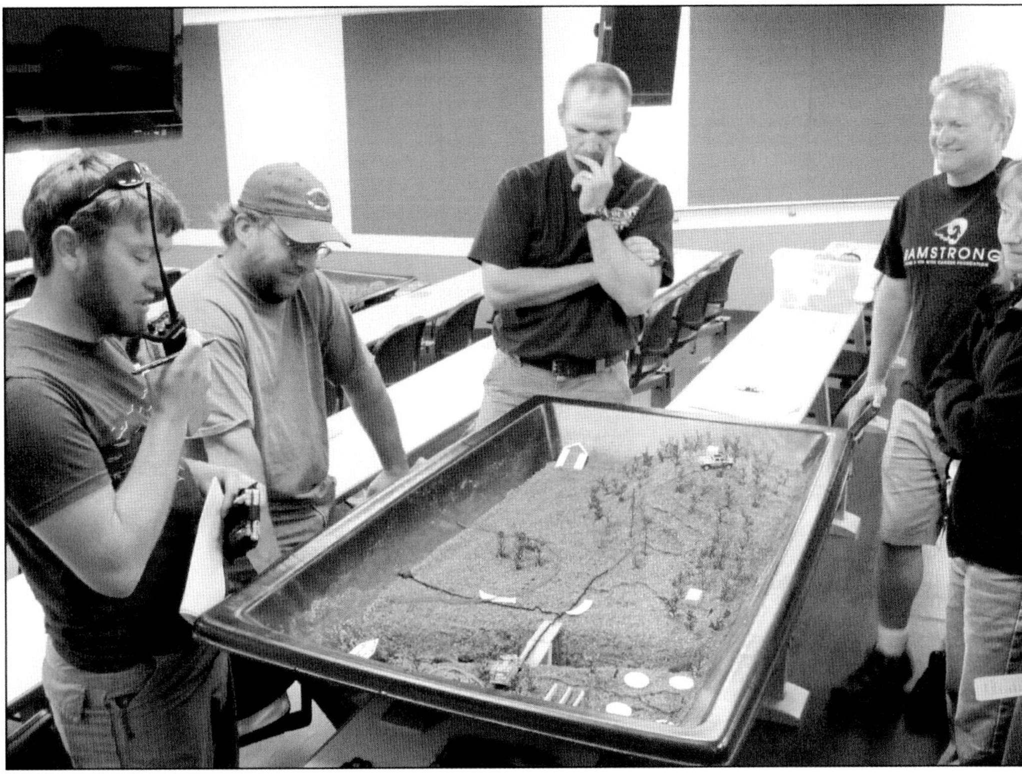

Wildland fire leadership students participate in a hands-on sand table exercise as part of training provided by the Big Rivers Forest Fire Management Compact. Students engage in a tactical exercise during the training while being provided input over a two-way radio. (BRFFMC.)

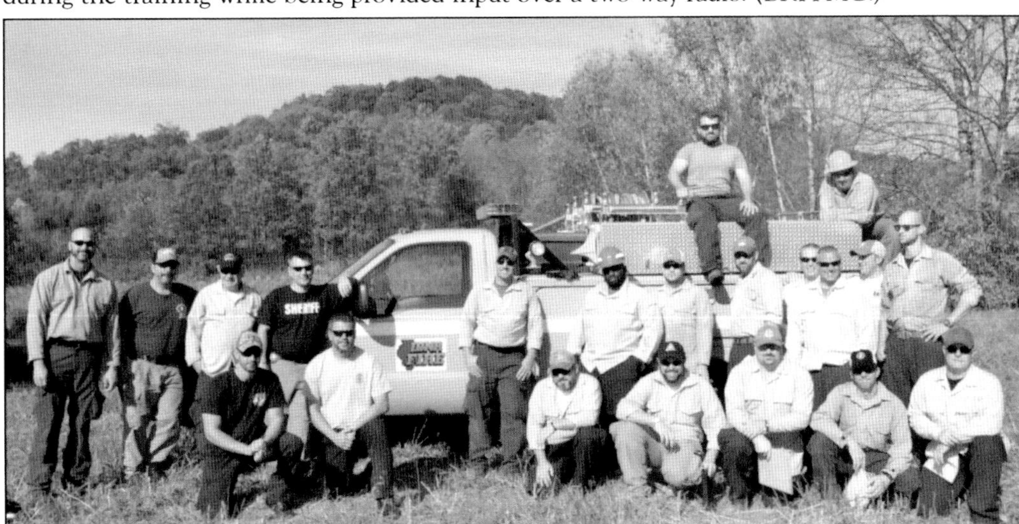

FI 210 Wildland Fire Investigation students pose for a picture during one of their hands-on field sessions in November 2016. The course was sponsored by the Big Rivers compact and provided the basic concepts of conducting an origin-and-cause investigation of wildfires. Wildland firefighters and law enforcement officers from county, state, and federal agencies participated in the training. An Illinois DNR Forestry Type 6 engine is behind the students. (BRFFMC.)

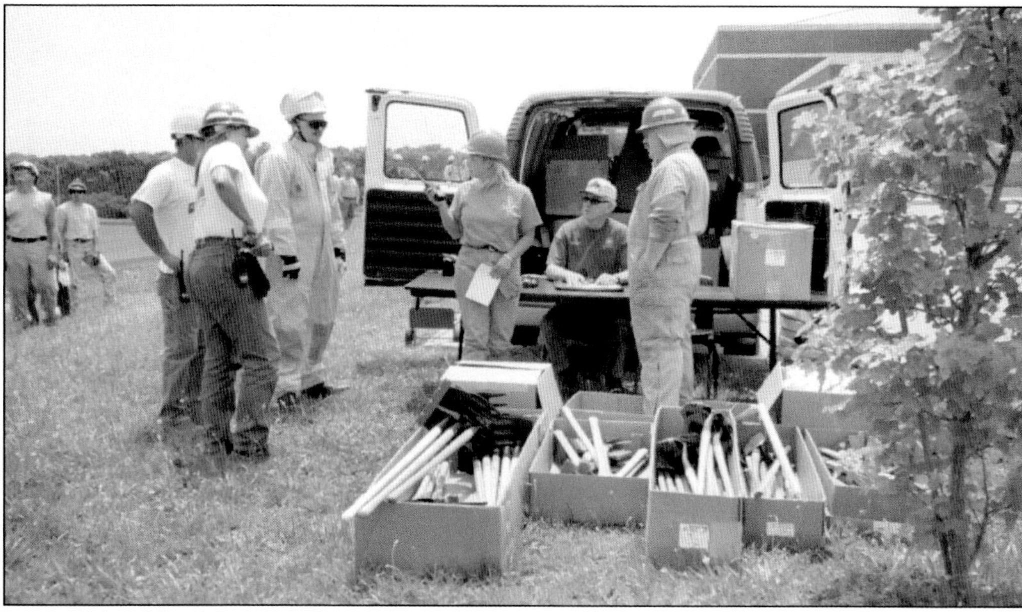

Basic wildland firefighting students receive their assigned hand tools from the academy cache during the 2003 Midwest Wildfire Training Academy. Note the boxes of Pulaskis, McLeods, shovels, and combi tools in cardboard boxes, which were on loan from the fire cache in Minnesota. (BRFFMC.)

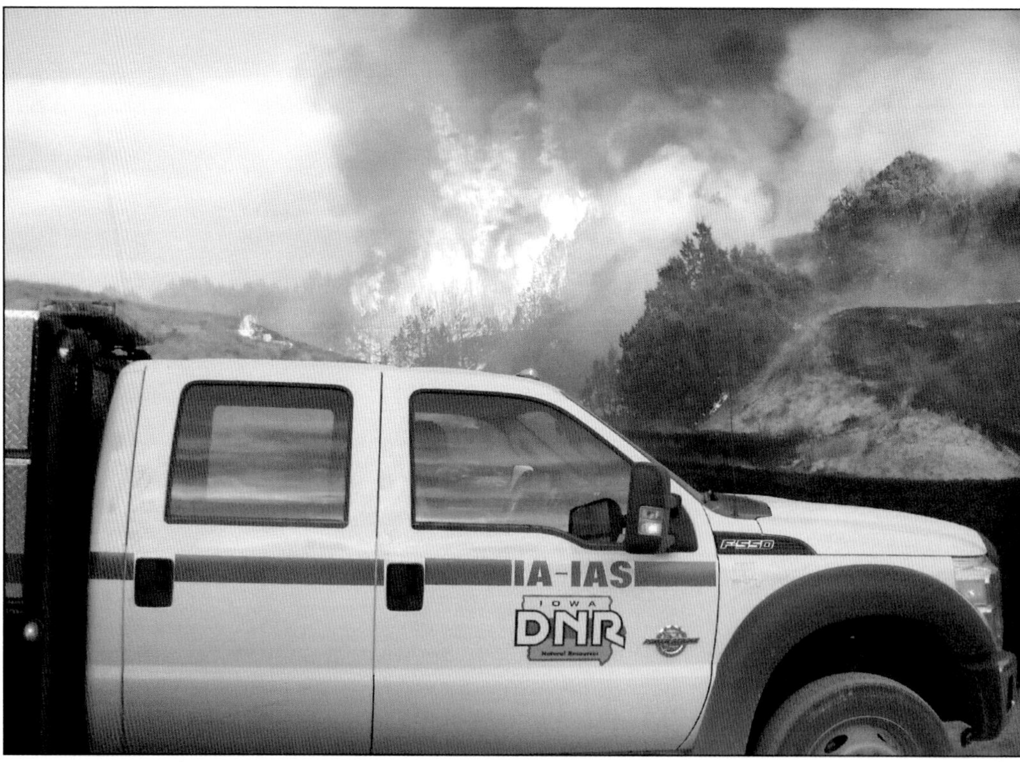

An Iowa Natural Resources Type 6 engine was assigned to a wildfire in South Dakota during the 2012 summer wildfire season. Annually, Iowa mobilizes wildland firefighters as part of an interagency team with agencies from within the Big Rivers states. (Iowa DNR Bureau of Forestry.)

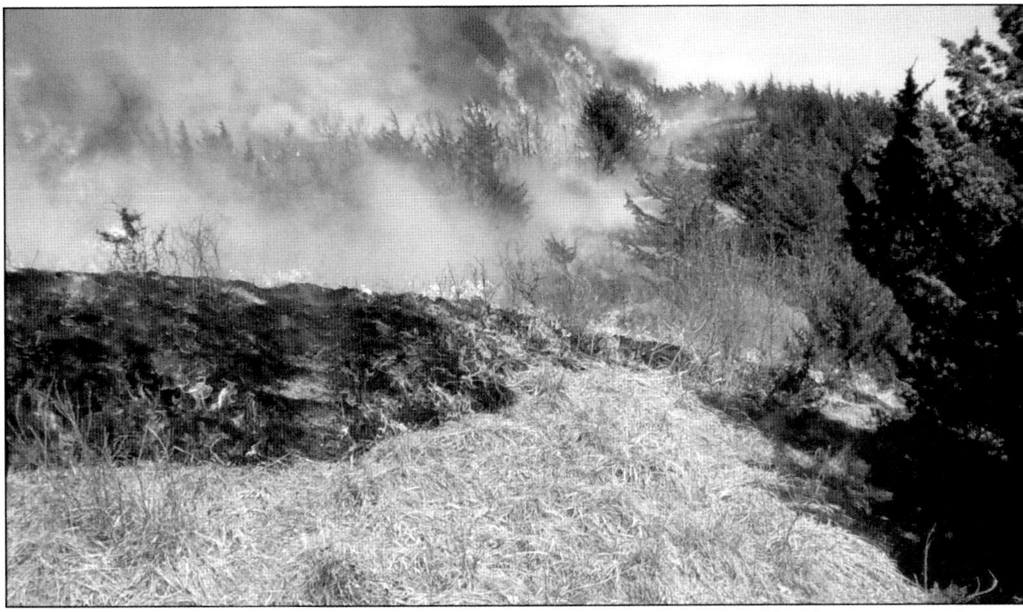

A fast-moving wildfire burns through Iowa's Loess Hills. Historically, the Loess Hills were characterized by prairie and bur oak savannas that were maintained by frequent fires ignited by lightning and Native Americans. During the 1800s and early 1900s, farmers burned the area to stimulate forage production and remove invasive plants. Today, prescribed fire is used as a management tool to improve wildlife habitat, forage, and oak regeneration and to manage encroaching weeds and brush. (Iowa DNR Bureau of Forestry.)

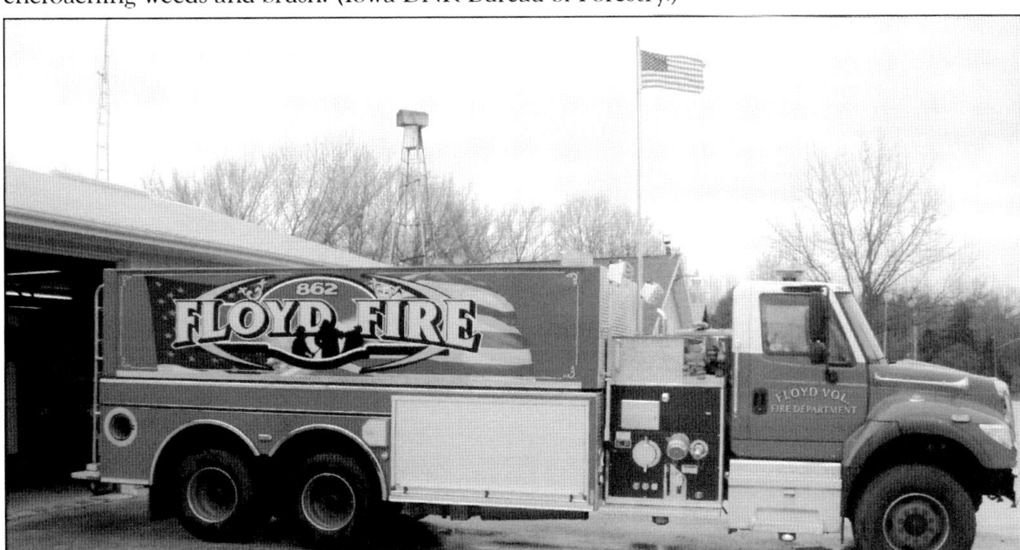

Iowa's Floyd Volunteer Fire Department received this tender through the US Forest Service's Department of Defense Firefighter Property program after refurbishing it through 3,200 hours of volunteer labor and local business donations. The truck was only a few years old but had been involved in a roll-over accident while owned by the Department of Defense. Floyd Volunteer Fire Department repaired the truck's tank, purchased and installed a new cab, and repainted the entire engine. It was placed in service in 2012. (Iowa DNR Bureau of Forestry.)

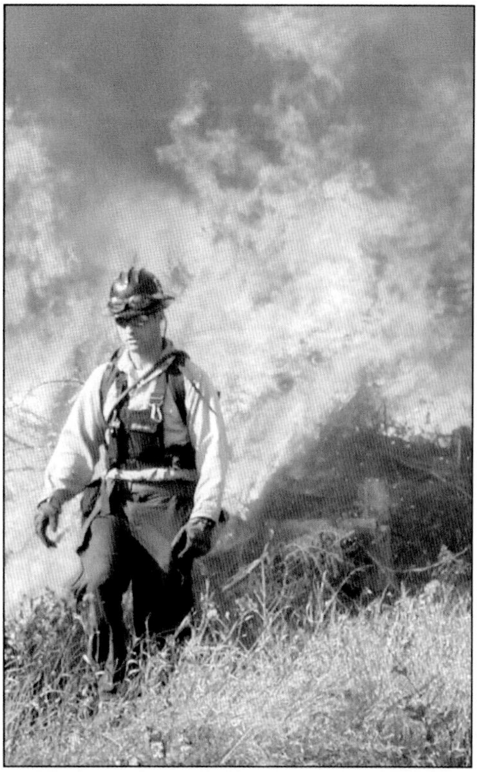

Pictured is Iowa DNR firefighter Chad Graeve operating at a prescribed fire project in 2011. Prescribed fire is an integral part of managing Iowa's tall grass prairies and savannas. In 2010, Iowa DNR finalized their statewide prescribed fire policy with the goal of using fire as a tool for ecological restoration and maintenance of natural areas on state-owned, -leased, or -managed lands and private lands where the DNR supports the use of fire as a management tool. In 2011, a total of 24,000 acres were treated with prescribed fire. (Iowa DNR Bureau of Forestry.)

The Virginia Township Fire Department in New Virginia, Iowa, operates this rural firefighting engine acquired through the Iowa DNR's Bureau of Forestry. Iowa Bureau of Forestry administers both the Federal Excess Personal Property program and Firefighter Property program that provides firefighting vehicles and equipment to fire departments across the state. This vehicle is a Stewart Stevens M1078 four-by-four light-utility truck, first used by the military in 1996. This unit has been converted for firefighting use. (Iowa DNR Bureau of Forestry.)

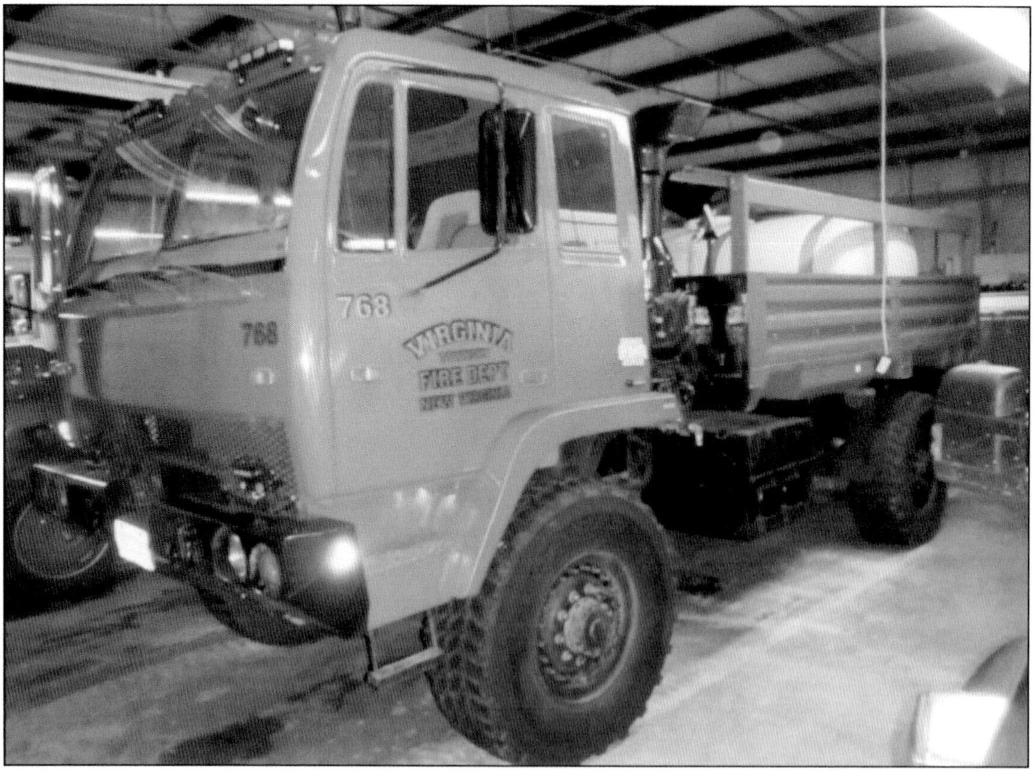

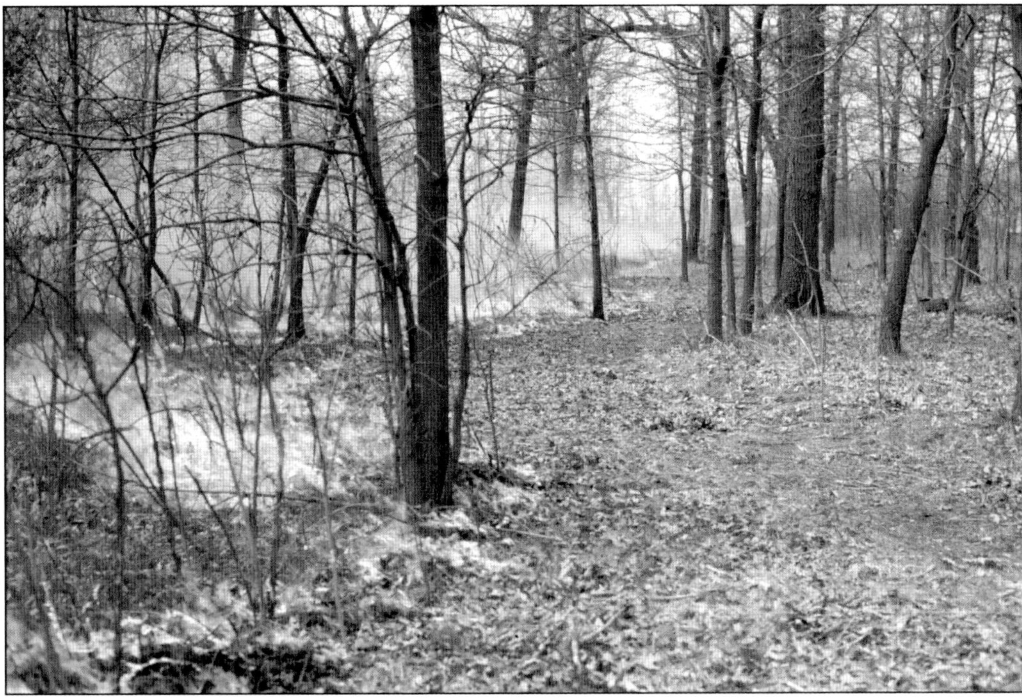

This wildfire is backing through an Iowa forest. In 2005, a total of 542 wildfires burned 20,925 acres across the state. Iowa's Bureau of Forestry's wildfire program is not mandated to suppress wildfires across the state, but it is responsible for providing wildland firefighter and prescribed fire training, supporting the use of prescribed fire, wildfire prevention, and the administration of the Federal Excess Personal Property and forest fire protection equipment programs, as well as the Volunteer Fire Assistance grant program. (Iowa DNR Bureau of Forestry.)

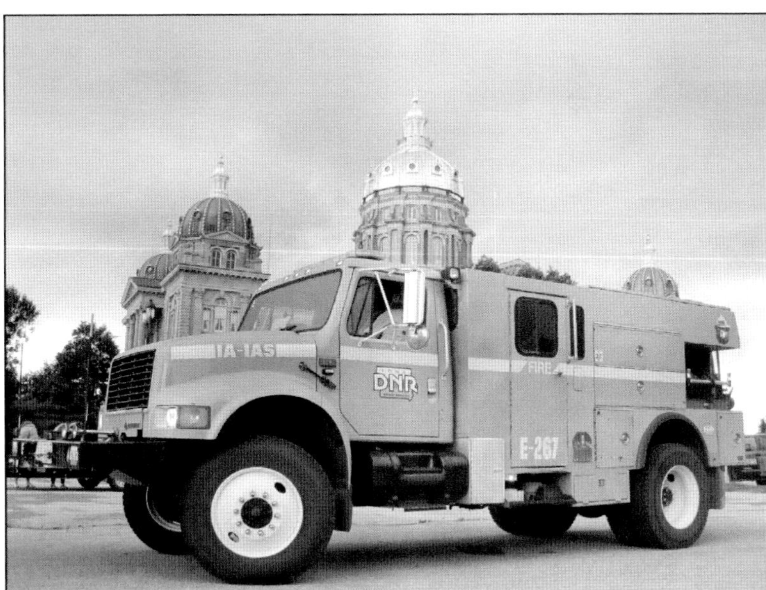

An Iowa DNR Forestry Type 3 engine participates in the 2016 Iowa State Fair Parade in Des Moines. The dome of the Iowa State Capitol is visible behind the engine. (Iowa DNR Bureau of Forestry.)

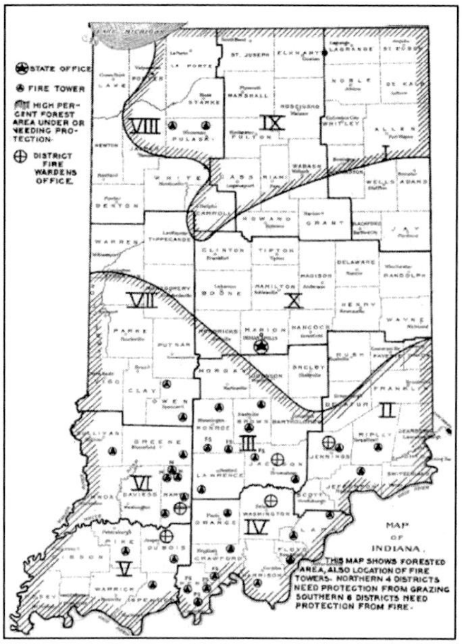

In 1929, Indiana's first state forester, Ralph F. Wilcox, developed a plan to build fire towers and introduce better methods of organizing fire crews. At the time, Indiana had only one fire tower, and the first challenges included determining high points on which to build new towers, as the state had not been completely mapped topographically. After 20 towers were erected, a complete visibility survey was done to determine that three towers needed to be moved and several had to be raised. Here is a 1952 map of Indiana's 32 lookout towers. (Indiana DNR Division of Forestry.)

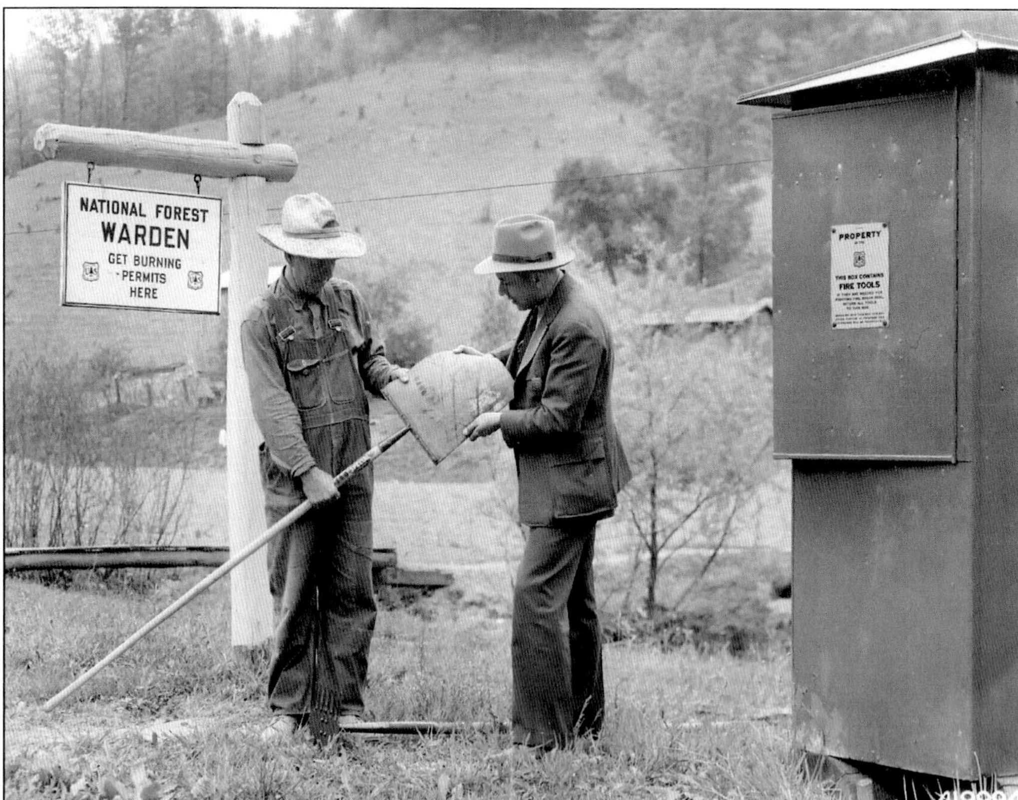

Both the Indiana Department of Conservation (now the Indiana Department of Natural Resources) and the US Forest Service employed fire wardens in the Hoosier state. The supervisor pictured is inspecting one of the US Forest Service fire warden's caches of tools, which were traditionally kept in bright-red buildings next to county roads. (Indiana DNR Division of Forestry.)

An unidentified Indiana tower man uses an alidade in the cab of his fire lookout tower to determine a bearing of a smoke from his location. The Osborne Fire Finder, the circular device in front of him, was a map with a mechanism that determined the position in the sky of a smoke. The fire's location could be pinned down when triangulated with other towers. (Indiana DNR Division of Forestry.)

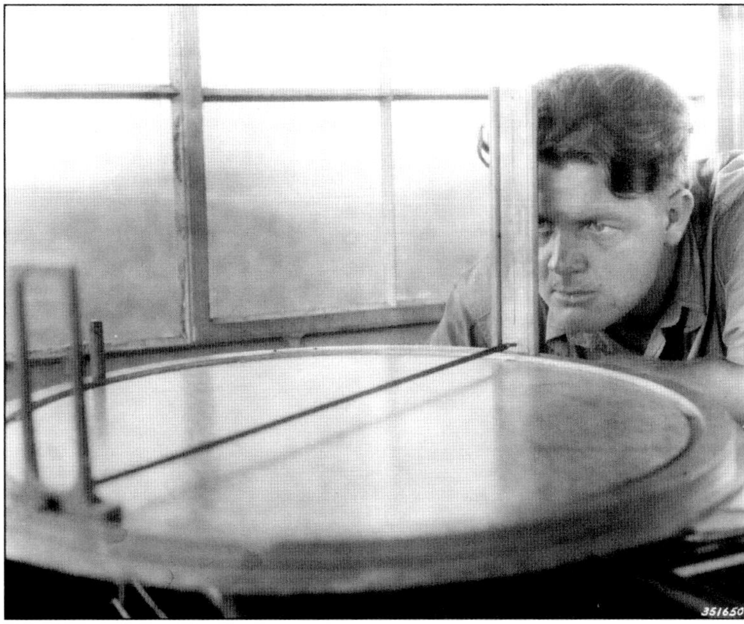

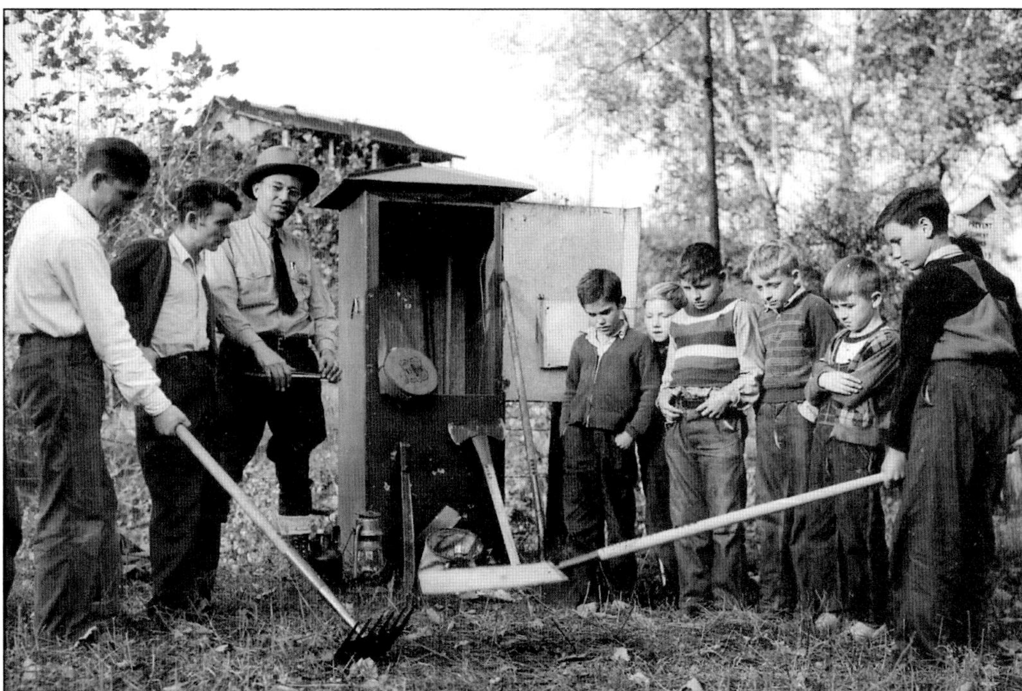

A group of boys is shown how to use a Kortick tool (now called a McLeod) to clear line and how to smother fires burning in grass with a "swatter." In 1959, the Indiana DNR Division of Forestry spent 8,654 hours educating their firefighters and the public, which included 2,400 boys that were enrolled in the Office of Civilian Defense's Forest Fire Fighters Service. High school boys were not paid to fight fires but were happy to get out of school. (Indiana DNR Division of Forestry.)

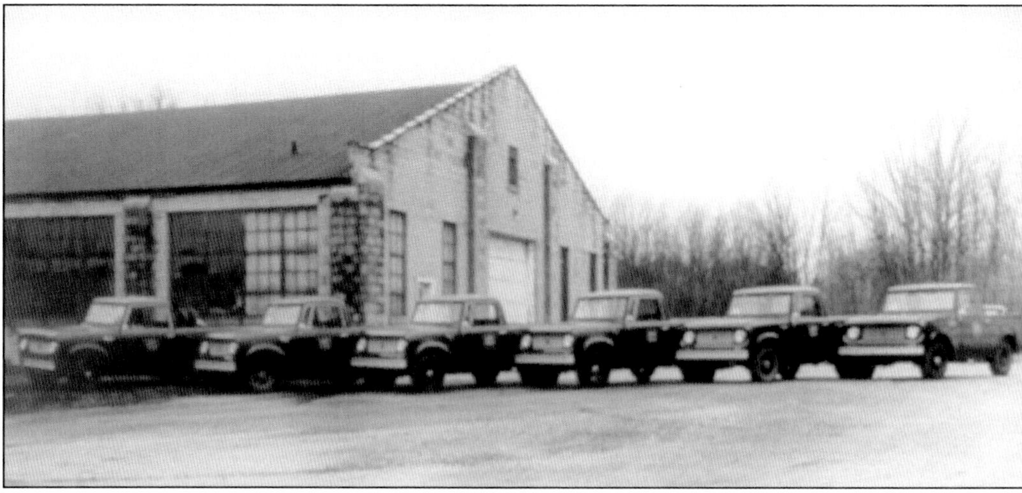

In 1964, the Indiana Division of Forestry established the position of fire coordinator to work directly with the Law Enforcement Division and other related agencies to take the lead in the state's fire control program. This photograph shows a lineup of new trucks to be outfitted for firefighting that year at the Indiana Fire headquarters' warehouse. (Indiana DNR Division of Forestry.)

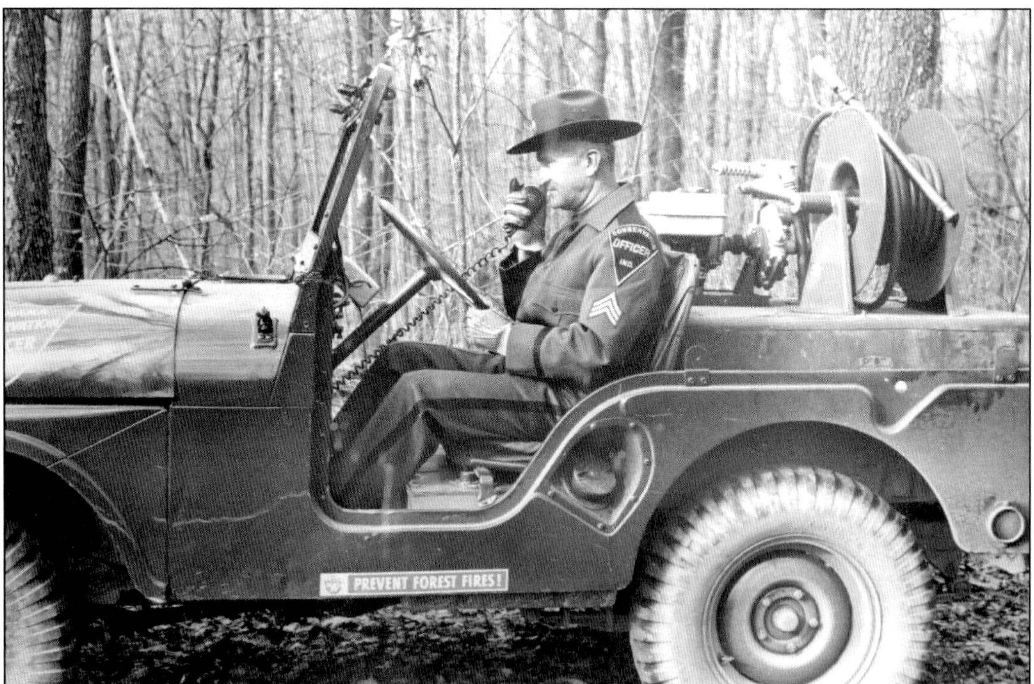

Pictured in 1963 is Roy White, quartermaster for the Indiana Department of Conservation. White received the jeep he is in from the Federal Excess Personal Property program and outfitted it for firefighting. The quartermaster's duties included managing the equipment program while serving in a dual capacity in law enforcement. He also inspected fire towers and served as an equipment operator, mechanic, and a communications specialist. (Indiana DNR Division of Forestry.)

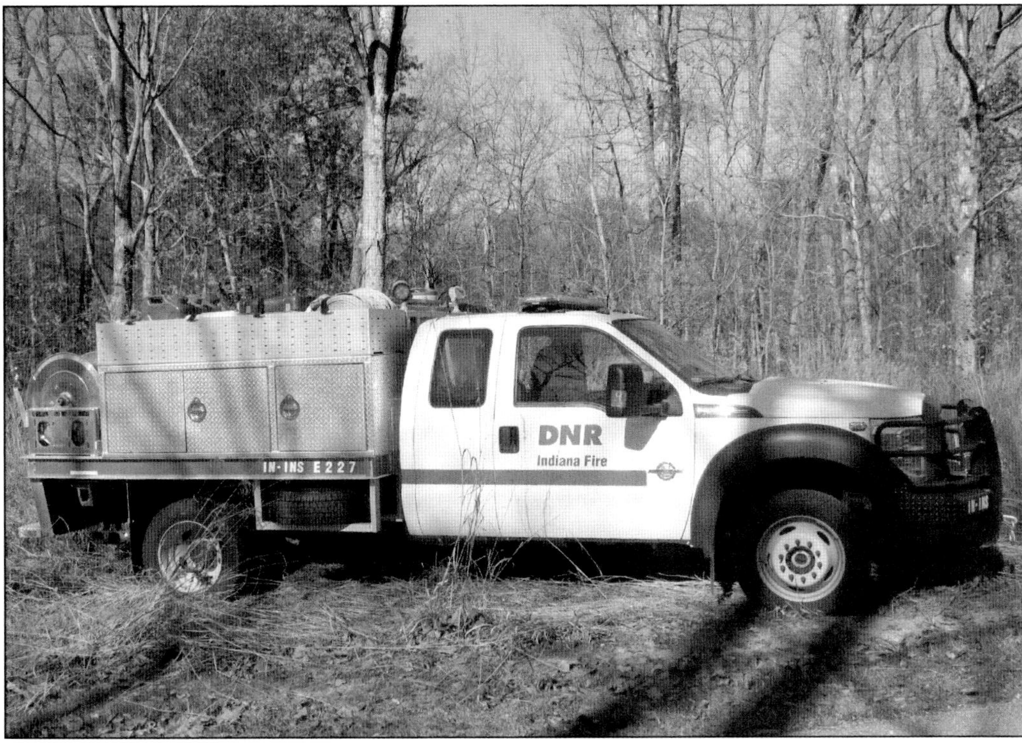

This 2016 photograph shows a modern-day Indiana DNR Division of Forestry Type 6 engine. This engine is outfitted with a tank, pump, and assorted fire hoses and can carry an assortment of hand tools, chain saws, safety gear, and miscellaneous firefighting supplies. (Indiana DNR Division of Forestry.)

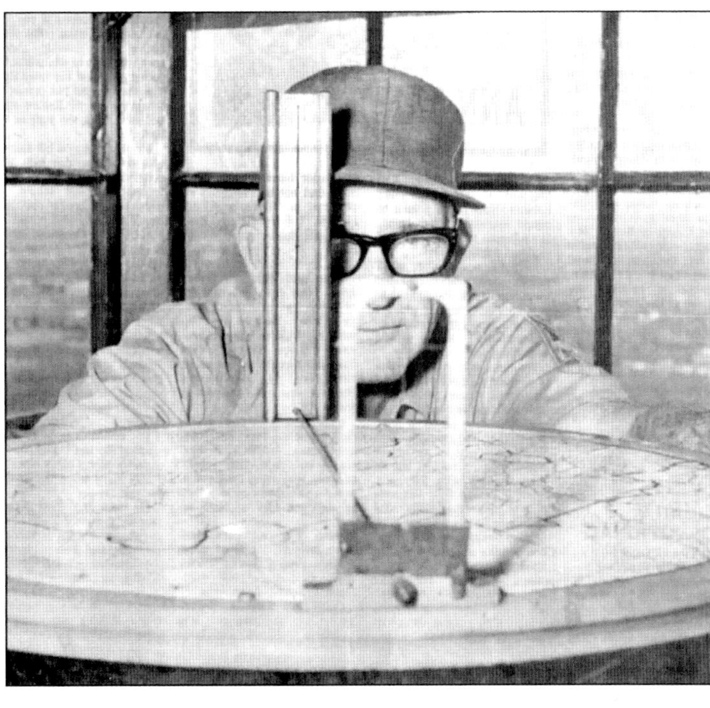

Missouri forest ranger Glenn Skinner puts his sight on smoke through his Osborne Fire Finder from the Knob Lick Lookout Tower. The Knob Lick Lookout Tower is located near Farmington, Missouri, in St. Francois County. The tower is still in use and has been opened to the public as a viewing platform. (Missouri DOC Forestry Division.)

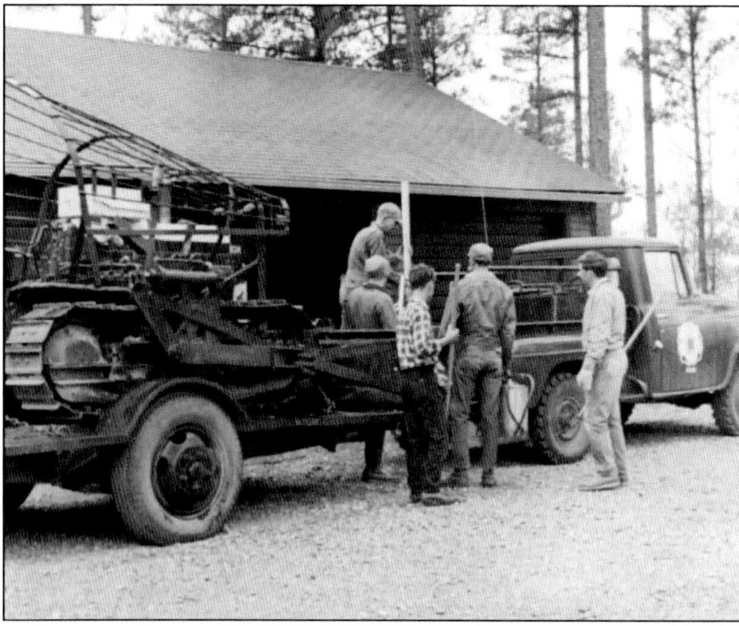

Bulldozers have been consistent pieces of the Missouri Department of Conservation's wildfire suppression program. This photograph was taken in 1965 at the Missouri Department of Conservation Eminence Field Office. This is a John Deere MC dozer, sitting on a Miller Tilt Top trailer being pulled by an International Power Wagon. Take note of the size of the dozer and the cage that was used to protect the operator. (Missouri DOC Forestry Division.)

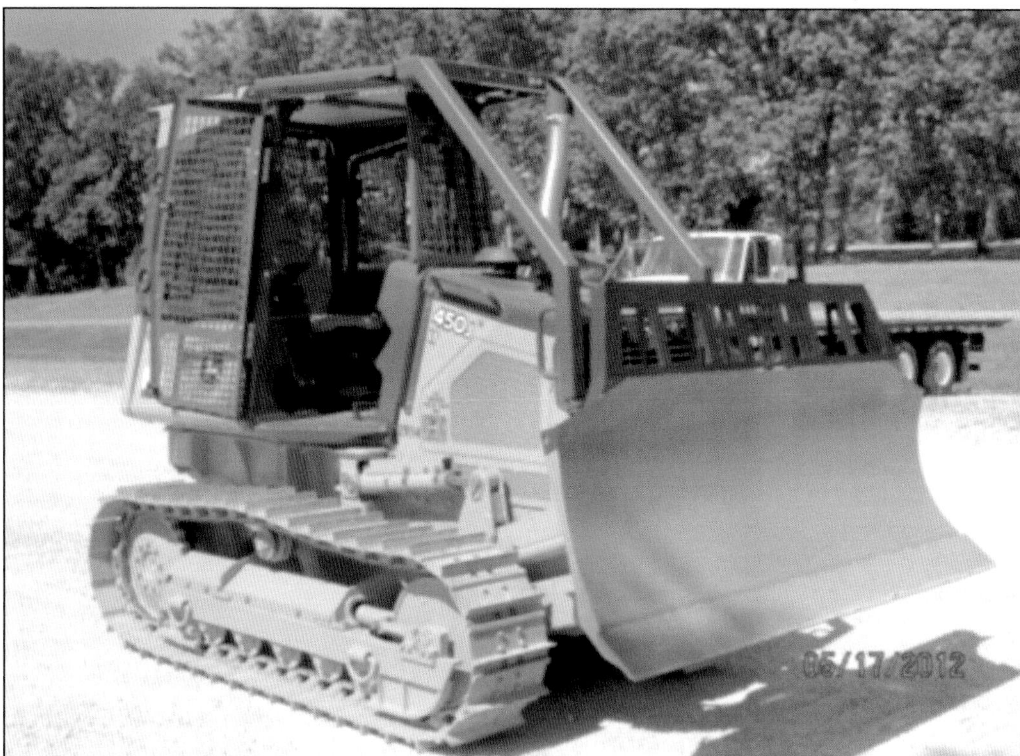

This fire dozer is the current version of the dozers used in earlier times within the Missouri Department of Conservation. Technology advances have produced a more reliable and safer piece of wildfire suppression equipment. The Missouri Department of Conservation operates a fleet of John Deere 450 wildfire suppression dozers, most of which are transported on International rollback trucks. (Missouri DOC Forestry Division.)

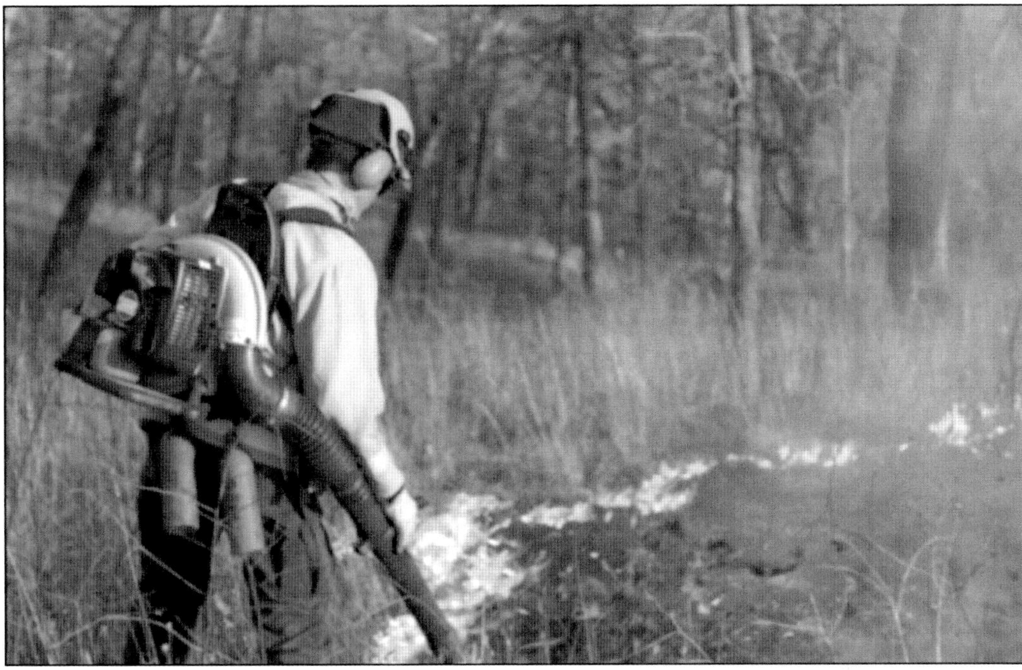

Leaf blowers continue to be a primary tool for fighting wildfires in the oak hickory forests of Missouri. Leaf blowers are used to clear leaf litter and small twigs and branches from a fire line. Advancements in equipment design over time have continued to improve firefighter suppression efforts. (Missouri DOC Forestry Division.)

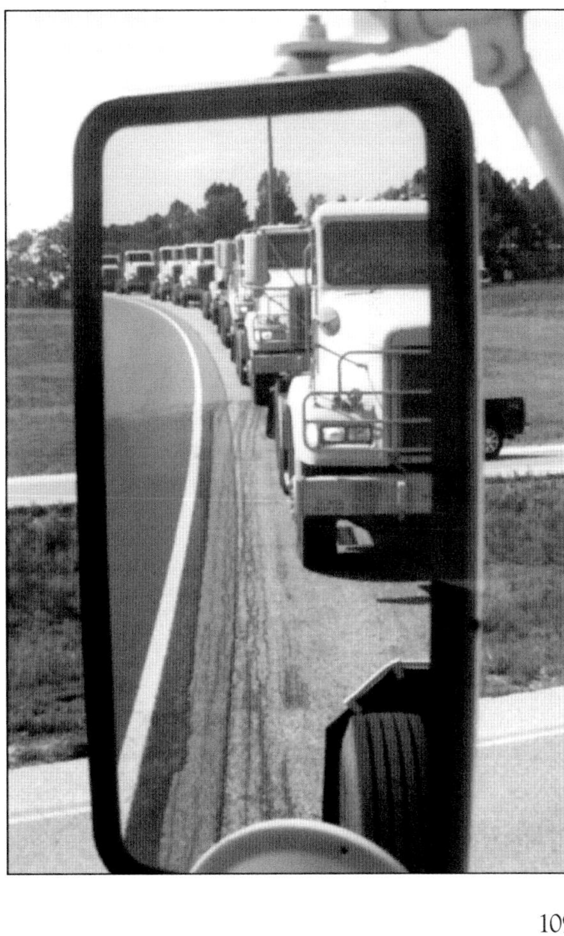

Twenty-nine Freightliner road tractors (Military M915A2 trucks) are being convoyed back to Lebanon, Missouri, from a military base in Nashville, Tennessee. Rural fire departments are converting these vehicles into 2,500-to-3,000-gallon water tankers to improve water supplies. The Missouri Department of Conservation has been acquiring property through the Firefighter Property program since August 2007. The program has proven to be invaluable for rural Missouri fire departments. (Missouri DOC Forestry Division.)

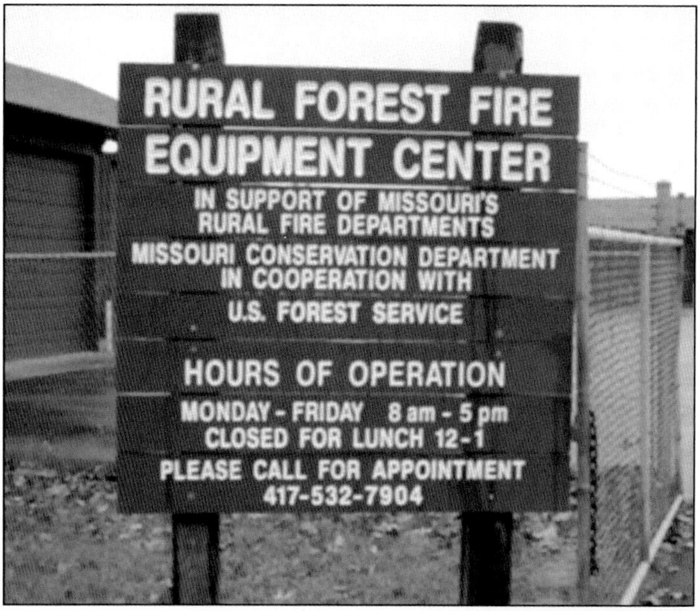

The Missouri Department of Conservation actively participates in both of the Federal Excess Property programs offered by the US Forest Service. Missouri Department of Conservation has been active in the Federal Excess Personal Property program since 1957 and became active in the Firefighter Property program in 2007. The Missouri Department of Conservation has been able to provide equipment to over 750 of the approximate 850 fire departments across Missouri. (Missouri DOC Forestry Division.)

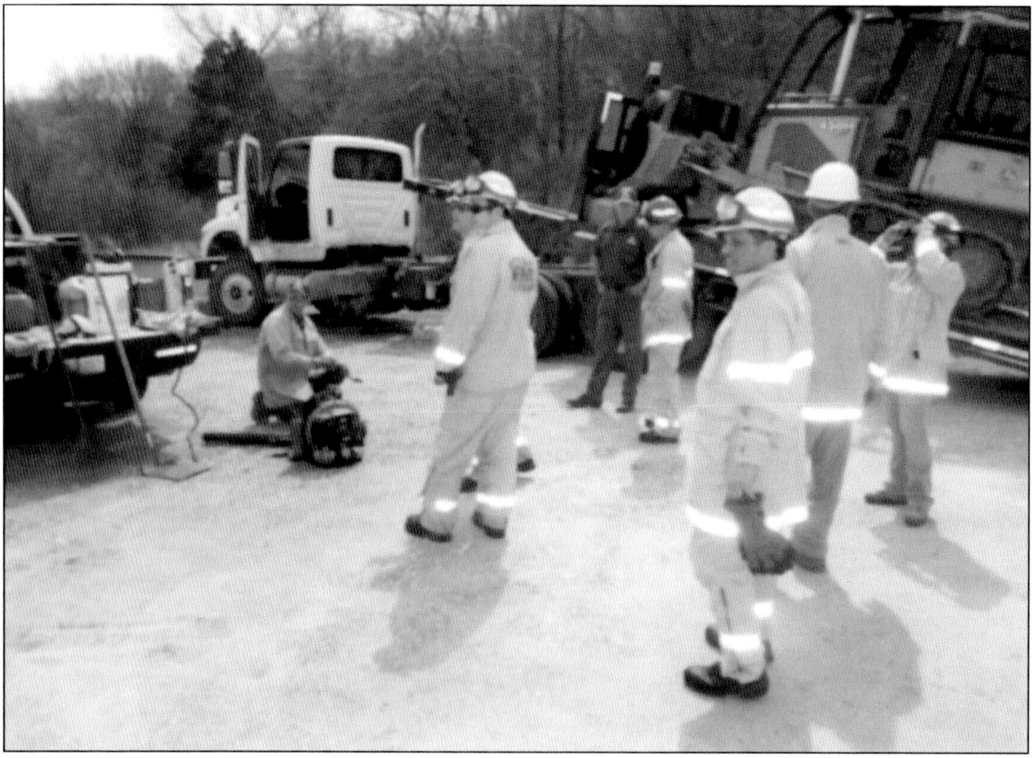

Missouri Department of Conservation staff members are very involved with the over 750 volunteer fire departments located across Missouri and actively provides wildland fire training to fire department staff. The Missouri Department of Conservation also provides 50/50 Volunteer Fire Assistance grants to fire departments. Funds for this grant are provided through the Missouri Department of Conservation from the US Forest Service. Funds are used to purchase wildland fire suppression equipment and communications equipment. (Missouri DOC Forestry Division.)

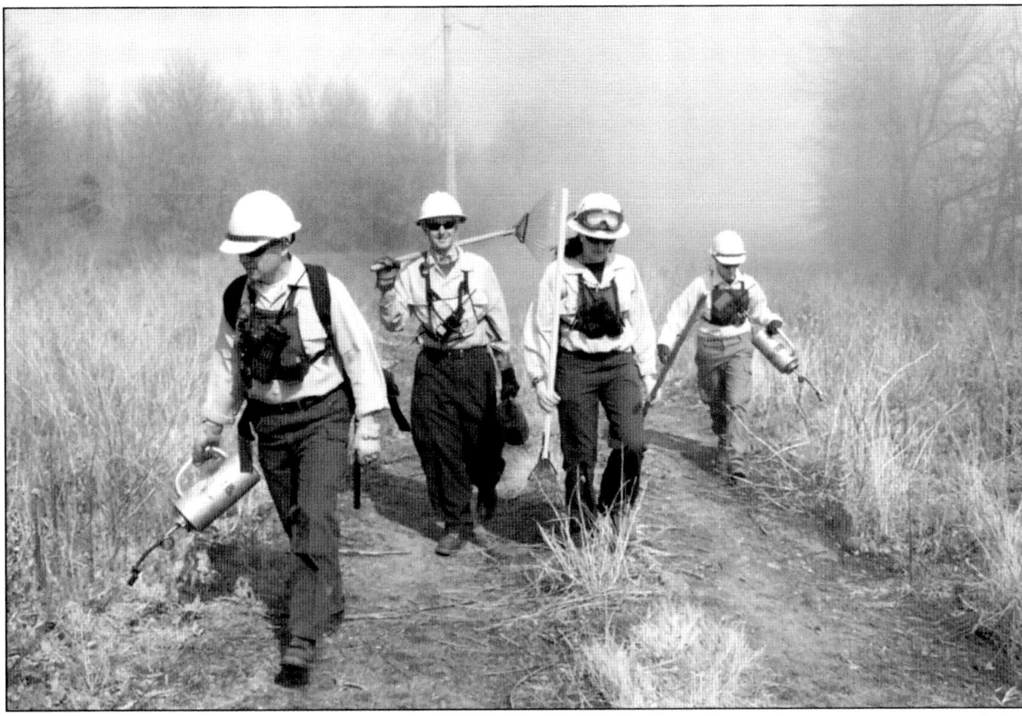

The Missouri Department of Conservation maintains a roster of approximately 450–500 qualified wildland fire personnel. These staff members are trained and maintain qualifications in wildfire suppression and the use of prescribed fire. In addition to maintaining a Missouri standard, many of these staff members also maintain federal wildfire qualifications and are mobilized throughout the summer months as wildfires occur across the country. (Missouri DOC Forestry Division.)

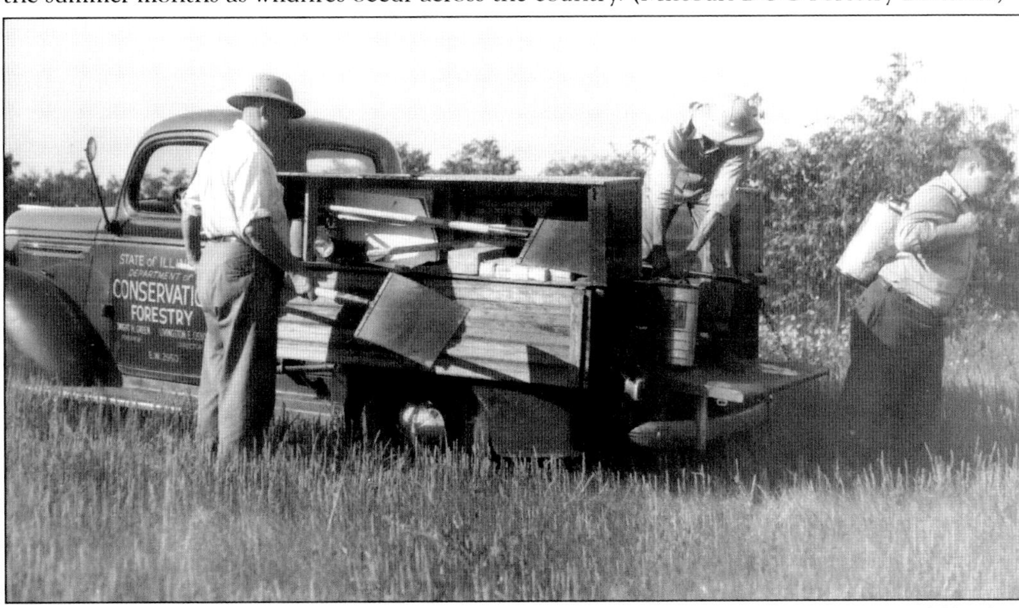

This photograph captures an Illinois fire crew preparing to conduct an initial attack on a wildfire from a 1960s Type 6 engine. Firefighters don firefighting backpack tanks as the engine boss checks out the assortment of hand tools available for their use. (Illinois DNR Division of Forestry Resources.)

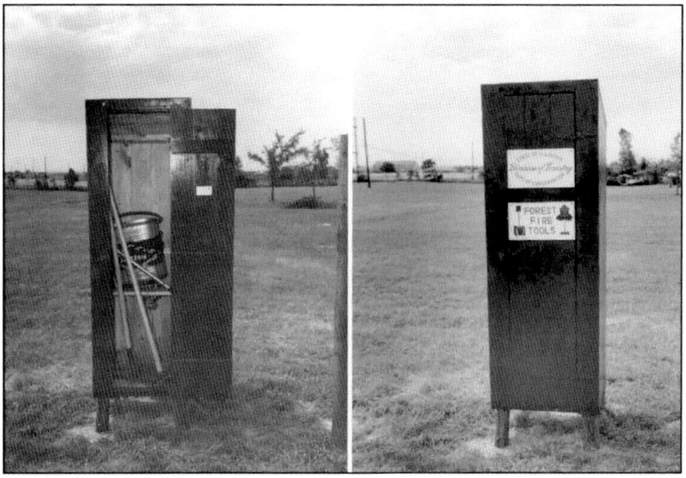

From the 1950s through the 1980s, the Illinois DNR Division of Forestry Resources maintained a network of roadside firefighting tool caches around the state. The caches held multiple flappers, shovels, Pulaskis, backpack pumps, and leaf and council rakes. These tools were available for passing travelers to utilize in containing roadside fires until trained firefighters could arrive. (Illinois DNR Division of Forestry Resources.)

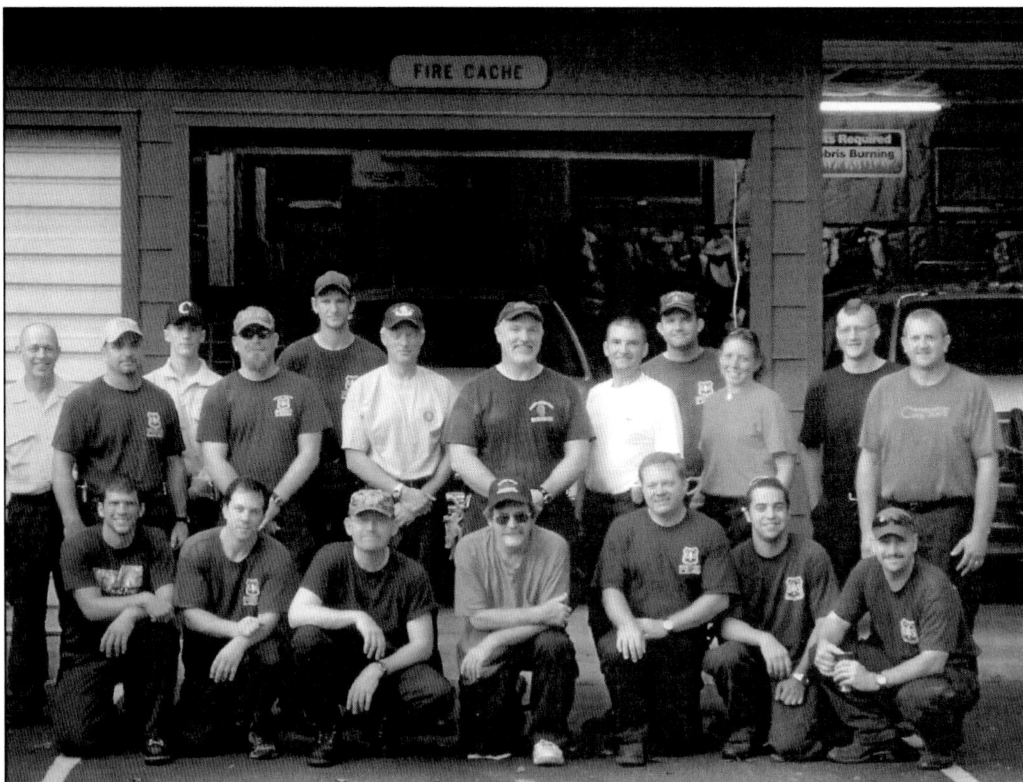

During the summer of 2006, Illinois mobilized its first out-of-state interagency fire crew with Illinois DNR personnel to an assignment in Minnesota. The Illinois DNR Division of Forestry Resources signed a cooperative mutual aid agreement with the US Forest Service to enable mobilizations in 2003. The first DNR staff member to mobilize was in 2005 as part of a Type 6 engine crew to Arizona. Since 2006, Illinois has mobilized at least one initial attack crew annually. (Illinois DNR Division of Forestry Resources.)

During a 1966 Illinois DNR Division of Forestry Resources staff meeting, Smokey Bear presented firefighting awards to staff members. This vintage Smokey Bear costume was outfitted with an Illinois DNR Division of Forestry Resources helmet in place of the campaign-style hat. (Illinois DNR Division of Forestry Resources.)

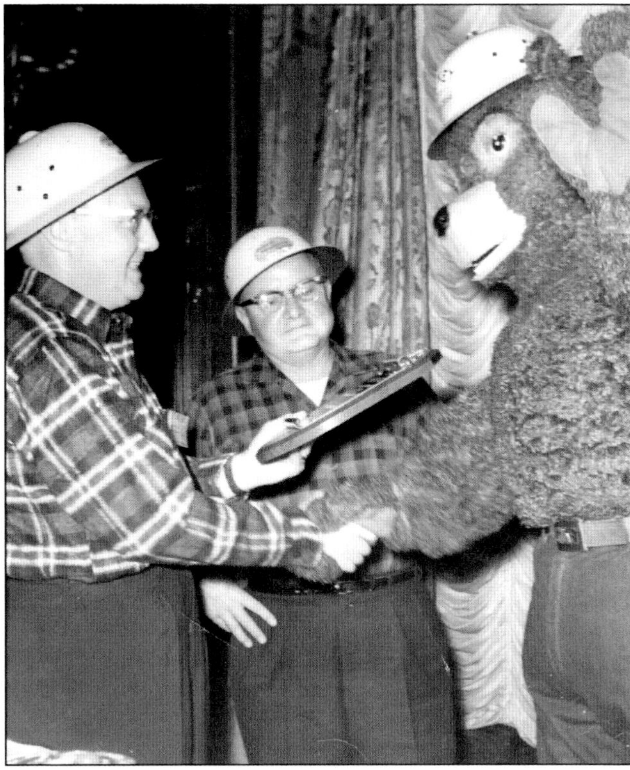

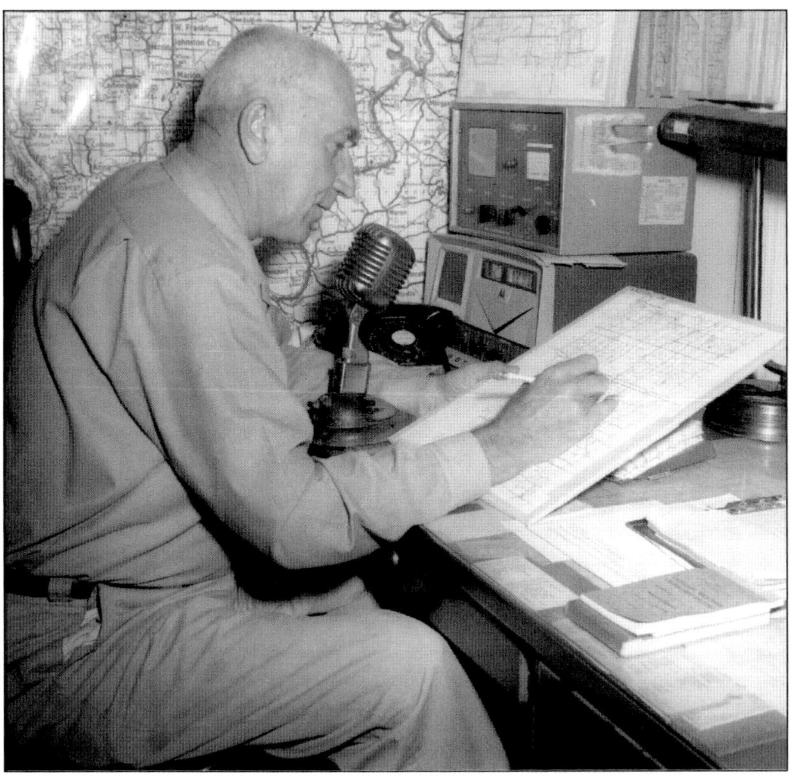

From 1950 through 1970, the Illinois Department of Conservation, Division of Forestry maintained an active wildfire fire suppression program. This picture shows forestry protection manager Max Lane dispatching firefighters to a wildfire in Southern Illinois while plotting the fire's location on a topographic map at Benton Fire headquarters during the 1966 fire season. (Illinois DNR Division of Forestry Resources.)

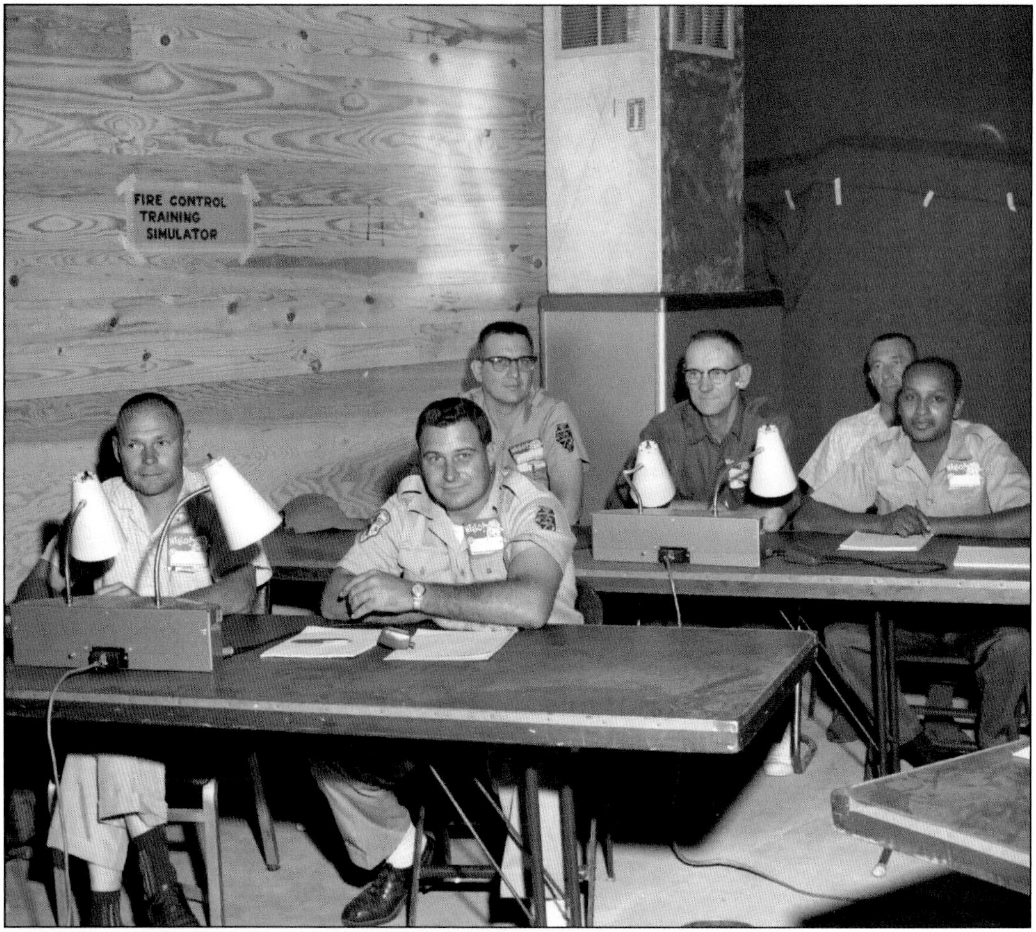

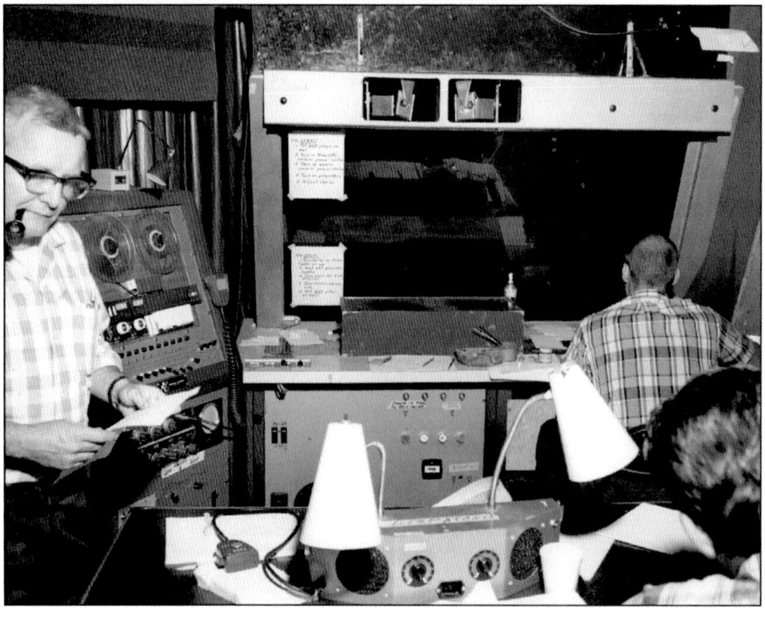

During the 1960s, all Illinois forestry staff participated in fire simulator training and were expected to be involved in wildfire suppression as part of their normal daily duties when required. These 1969 photographs capture a group of students participating in a simulation exercise above and a behind-the-scenes simulator operation at left. (Both, Illinois DNR Division of Forestry Resources.)

Six

Roscommon and Equipment

Wildland firefighters have been innovative in the tactics they have used to fight fires, including the acquisition, development, and construction of equipment and vehicles to support their efforts. Early vehicles were used to carry equipment and firefighters, and off-road engines and tractor plows were used to help in the direct attack of fires. Across the northeastern area with varied vegetation, fire behavior, and fire protection concerns, many specialized vehicles have been put to work over the years. Vehicles and heavy equipment were outfitted to handle their specific local jobs. The pine barrens in New Jersey, Long Island, and Massachusetts were equipped with brush trucks, stump jumpers, and breakers while the Great Lakes had bombardiers, engines, and crawler pulled fire line plows. Four-wheel drive engines, leaf blowers, drip torches, and specialized hand tools, such as leaf rakes, were used across all 20 states.

In June 1971, the Northeastern Forest Fire Control Supervisors agreed to accept a proposal from the State of Michigan to share the facilities of the Michigan Department of Natural Resources' Forest Fire Experiment Station in Roscommon, Michigan. At the subsequent meeting of the Northeastern Area State Foresters, this proposal was approved on a three-year trial basis with an annual estimated budget of $30,000. The Roscommon Equipment Center (REC) was fully staffed and in operation under the direct supervision of Steve Such by January 1972. John Demars, a maintenance mechanic, and Richard Greenlaw, a draftsman, were hired as part of the program. Direction for REC's work was from the Equipment Development and Testing Committee through the chief of the Michigan Forest Fire Division. This committee initially presented a list of eight high-priority projects for development and research at REC. The initial projects, among others, included electrical wiring conversions on military vehicles and evaluation of trucks, tractors, plows, foam, and bombardiers. Creating vehicle schematics and specifications, protective tractor canopies, air-cooled safety helmets, and rustproofing treatments for water tanks were additional projects.

In 1999, REC became a National Equipment Center and has received funding from the US Forest Service on behalf of the National Association of State Foresters and all state forest fire management agencies within the United States. Since its creation, REC has developed over 70 projects and 20 news-notes related to forest fire management and firefighting equipment. REC has assisted countless agencies and fire equipment specialists with technical support.

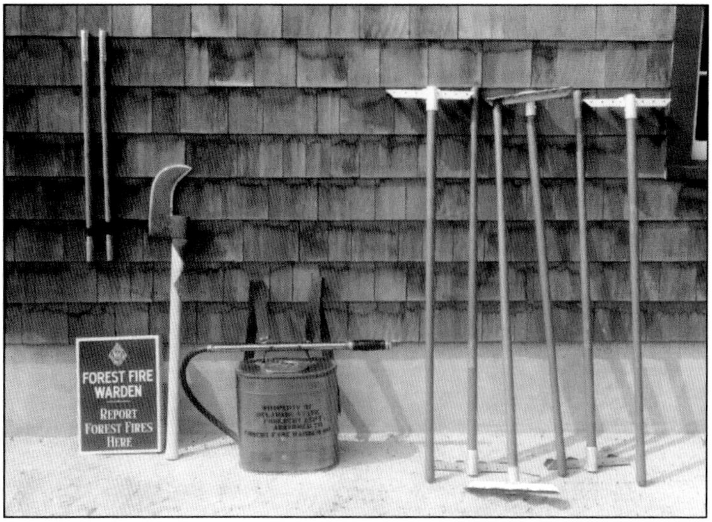

Prior to mechanization of firefighting, hand tools were the primary equipment utilized to combat wildfires. This photograph shows the full complement of tools issued to each Delaware forest fire warden in 1941. Upon its establishment in 1927, the first state forester, William S. Taber, believed that forest fire control was the Delaware Forestry Department's "most important responsibility." (Delaware Forest Service.)

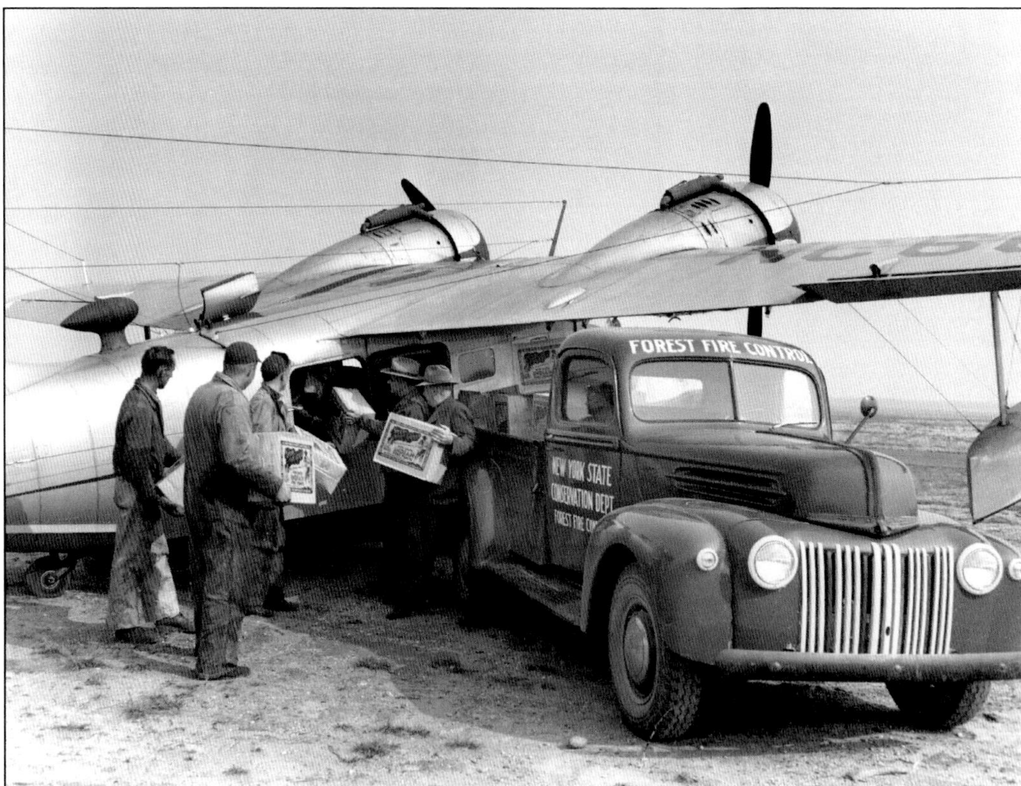

Over the years, many vehicles have been used to support wildland firefighting operations. This 1930s photograph shows New York forest rangers offloading supplies from a New York Forest Fire Control Ford pickup truck into a Grumman G-21 aircraft. As equipment has developed and improved, wildland firefighters have taken advantage of it whether from commercial enterprises or from military sources. (New York DEC State Forest Rangers.)

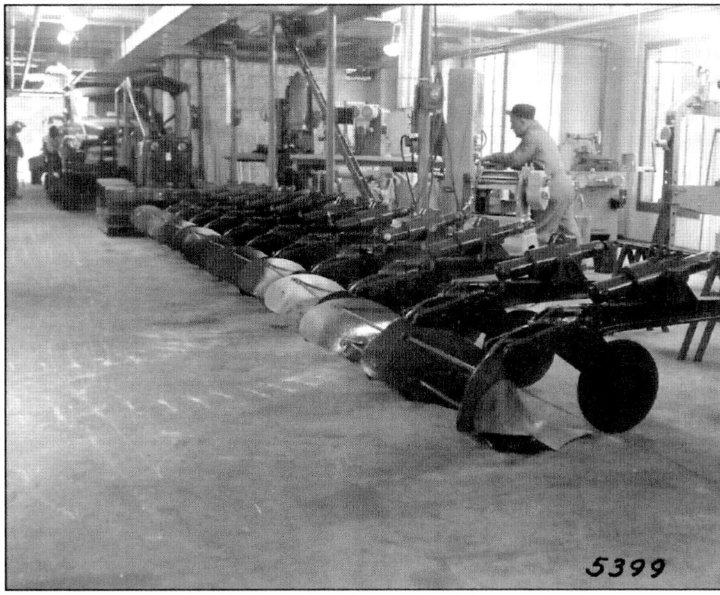

The Michigan Forest Fire Experiment Station was created in 1929 at Roscommon by the Michigan Department of Conservation in conjunction with the US Forest Service's Lake States Forest Experiment Station. This photograph shows an inside view of the original shop with a lineup of new fire line plows. The original buildings at the station were constructed by the Civilian Conservation Corps. (Michigan Roscommon Equipment Center.)

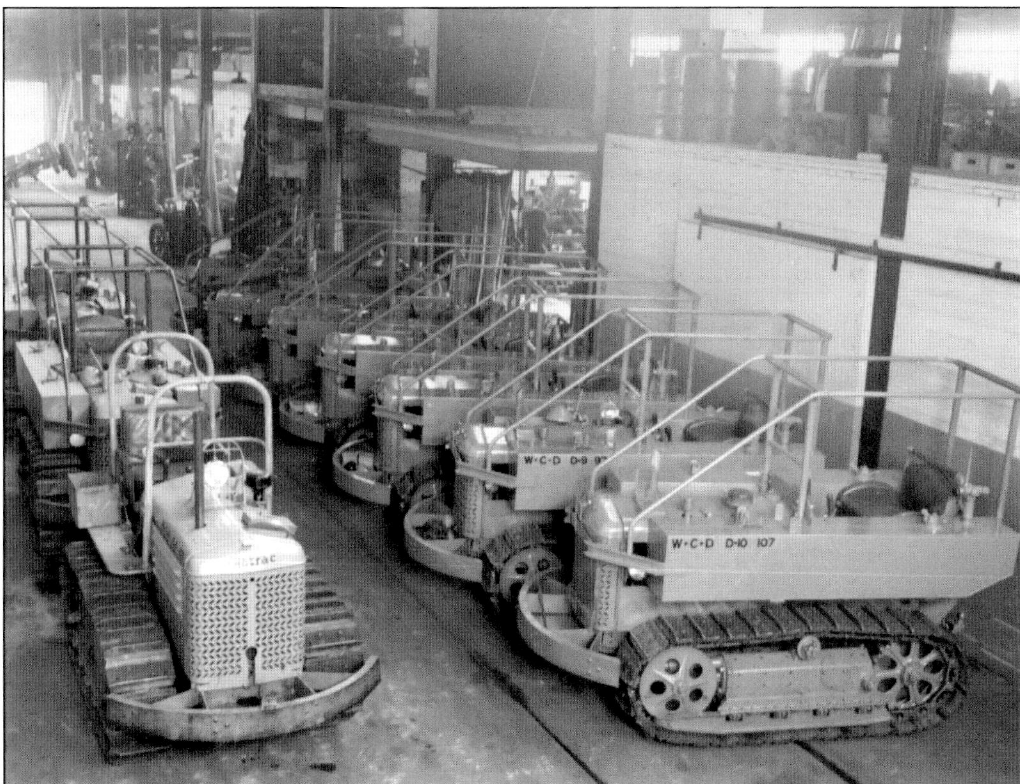

As equipment technology has evolved, wildland firefighters have taken advantage of it to develop and use specialized equipment to aid in the suppression of wildfires. This 1949 photograph captures a fleet of new Cletrac tractors inside the Wisconsin Forest Protection headquarters' equipment construction shop; the tractors are awaiting outfitting into fire line tractor plows at Tomahawk. (Wisconsin DNR Division of Forestry.)

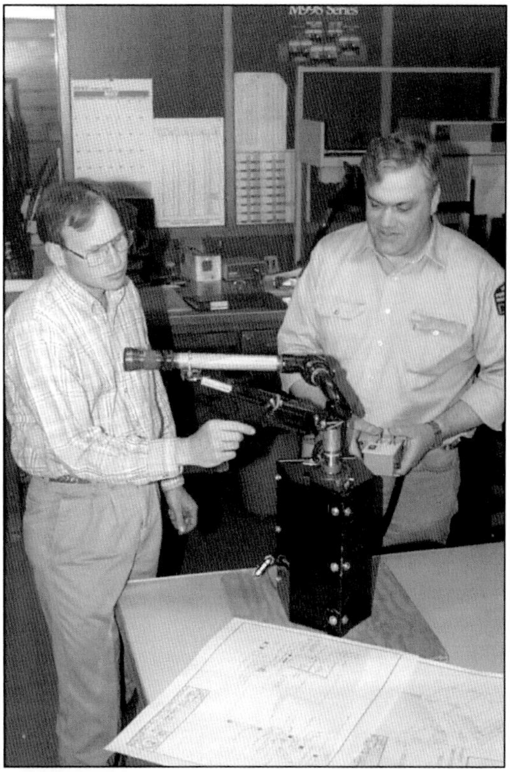

REC Project No. 58 developed drawings to produce a one-inch remote control water turret for use on wildland engines. The improved design was based from an "Aggie Roll 'N Squirt" turret, designed by the Texas A&M Engineering School, and was utilized by the Indiana DNR. In this photograph, from left to right, Brian Hutchins and Kirk Bradley (REC employees) compare the drawings to the actual turret. (Michigan Roscommon Equipment Center.)

This photograph captures Michigan DNR employees fabricating a body for a Michigan wildland engine at the Michigan Forest Fire Experiment Station. The Roscommon Equipment Center is part of the station. Over the years, the success of the firefighting equipment programs at all 20 northeastern states, as well as at REC, have depended on the fire managers, firefighters, engineers, and equipment maintenance and development personnel who develop, operate, and maintain the equipment daily. (Michigan Roscommon Equipment Center.)

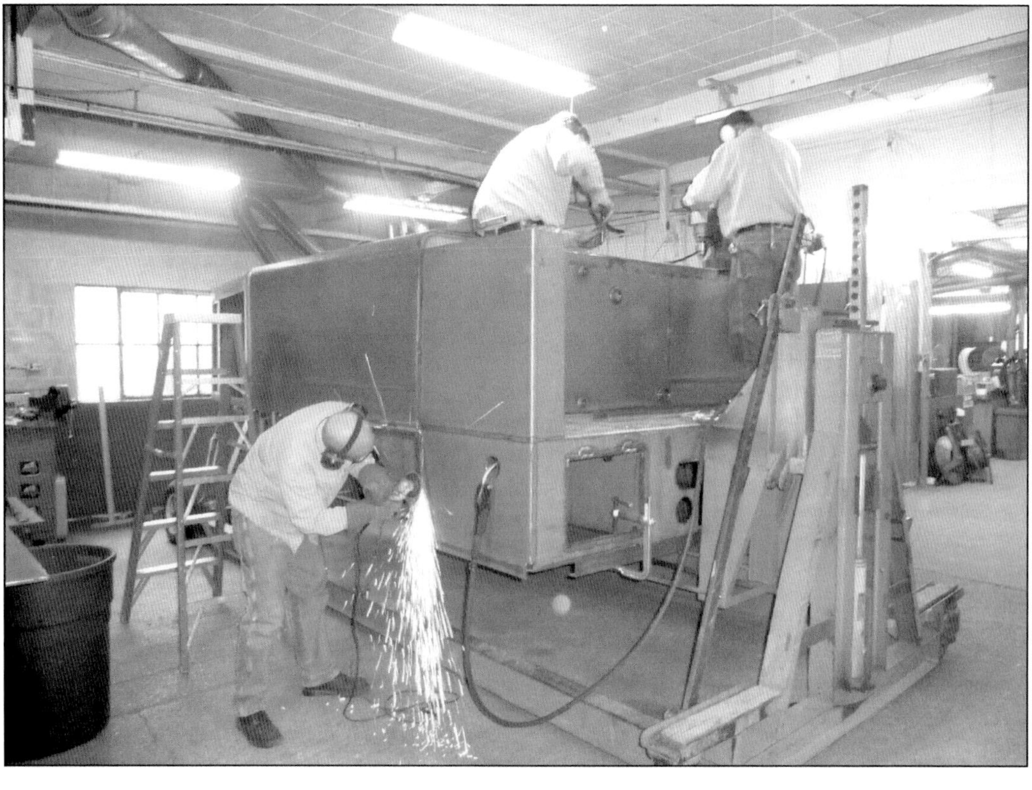

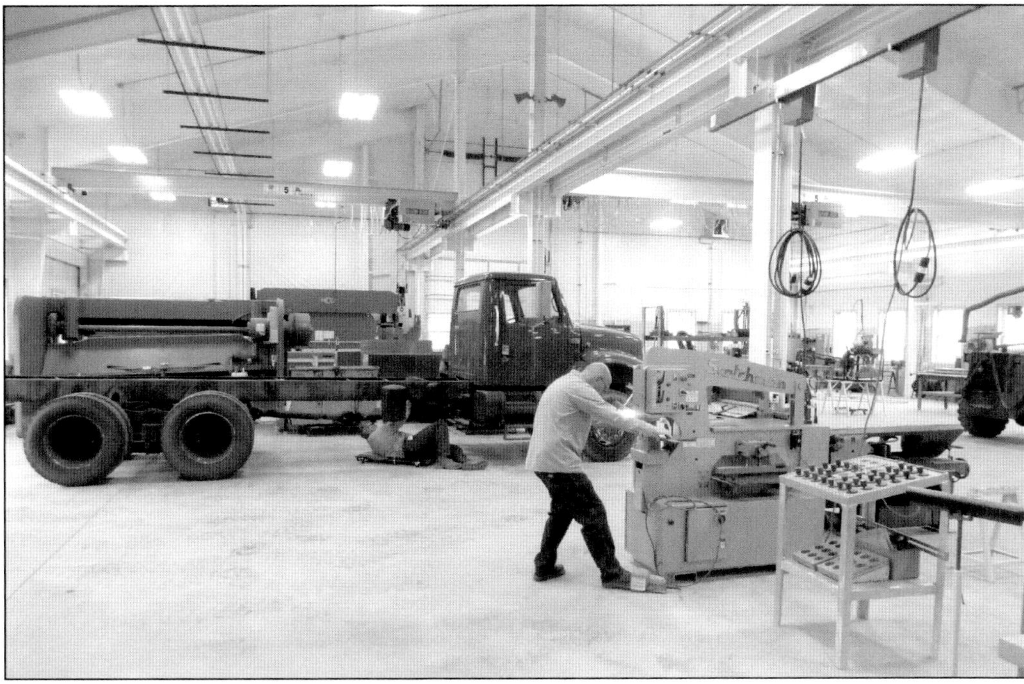

The Michigan Forest Fire Experiment Station moved into a new facility in 2014. This facility hosts the REC that supports all 50 states through an agreement with the US Forest Service's Northeastern Area State & Private Forestry. A forest fire simulator is also housed in the facility to train fire managers using virtual firefighting scenarios. This photograph shows a portion of the fabrication and assembly area. (Michigan Roscommon Equipment Center.)

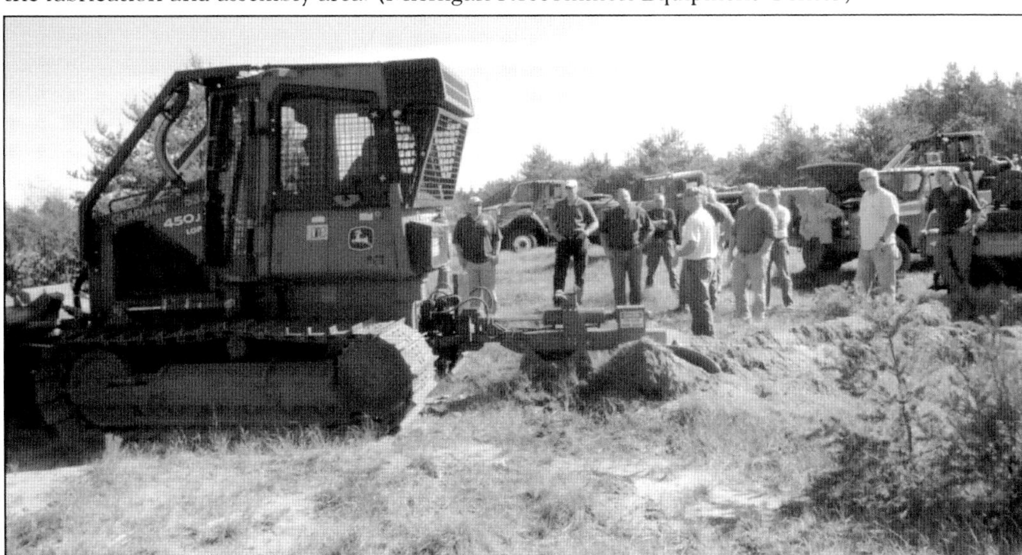

In this 2011 photograph, REC personnel demonstrate a JD 450 dozer with a Michigan-style fire line plow as part of a REC Mini Equipment Workshop. These workshops are held annually and participants come from state and federal wildland fire management agencies across the United States. Workshops are designed to demonstrate the latest wildland firefighting equipment and share ideas and innovations. (Michigan Roscommon Equipment Center.)

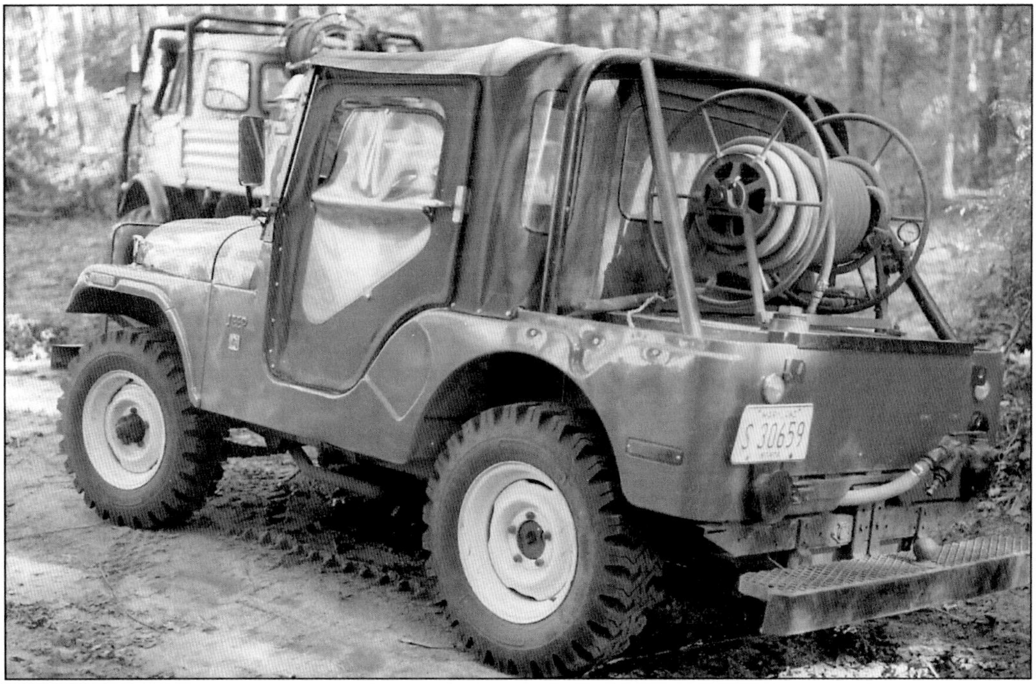

Jeep Tanker Handbook (Project No. 4) was developed as a result of the increased use of FEPP Jeeps in the Northeast. From 1973 through 1975, over 215 Jeeps were acquired through the FEPP program by the 20 northeastern states. Indiana, Delaware, Illinois, and West Virginia received prototype Jeep tankers from REC in 1973. Comments were solicited by all users during this period and were included in the final handbook with modification plans for use by states in converting Jeeps into firefighting units. (Maryland Forest Service.)

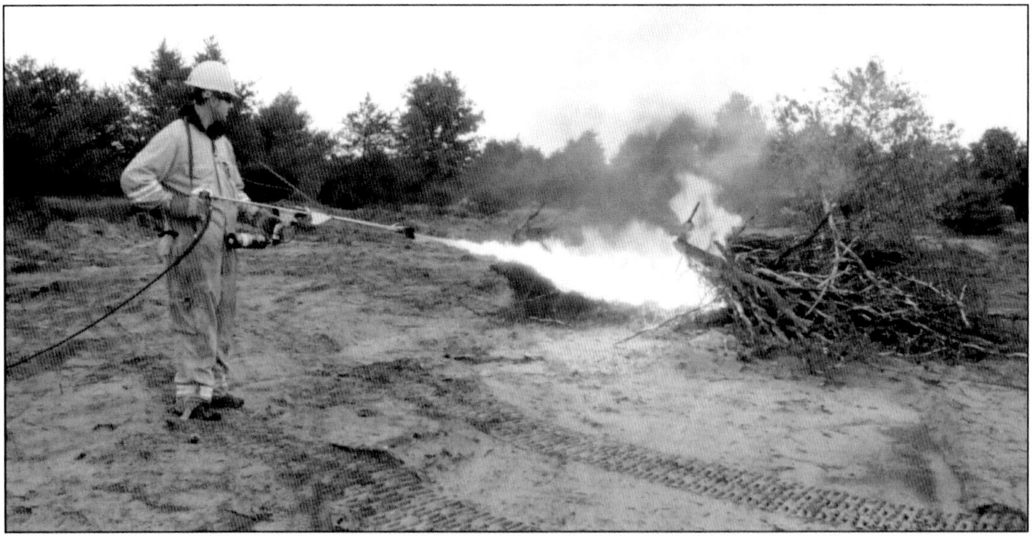

News-Note No. 18 (2014) outlines a flamethrower device that was developed with a hand-carried flamethrower wand accompanied with a slip-on type pumping unit that is carried in the cargo bed of a utility vehicle. The flamethrower device projects out standard drip torch fuel with the primary function of being used to ignite remote pockets of fuel some distance away from the user during prescribed burning operations. (Michigan Roscommon Equipment Center.)

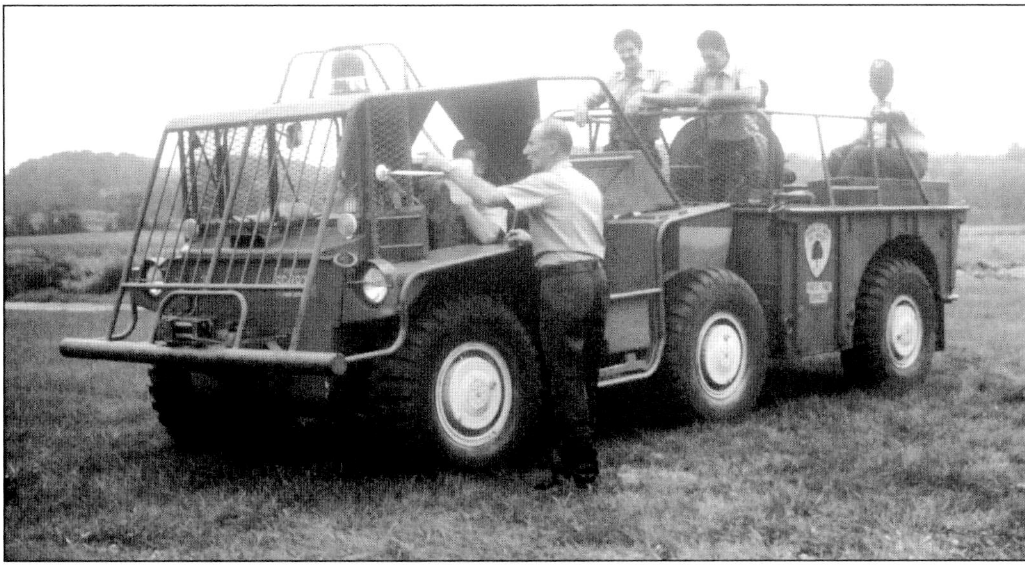

During the spring of 1988, Michigan, Minnesota, New Jersey, and Pennsylvania field tested the Gamma goat for firefighting use on over 45 wildland fires. These field evaluations were incorporated into REC Project No. 53 A and B, which outlined its best use and the design of a hard-cab and slip-on tank for the vehicle. Field testing found that the vehicle was best suited for rugged, rocky terrain and steep slopes. Bill Resch (New Jersey) directs the operator of the unit at a MAIFFPC meeting as Delaware firefighters ride on back. (New Jersey Forest Fire Service.)

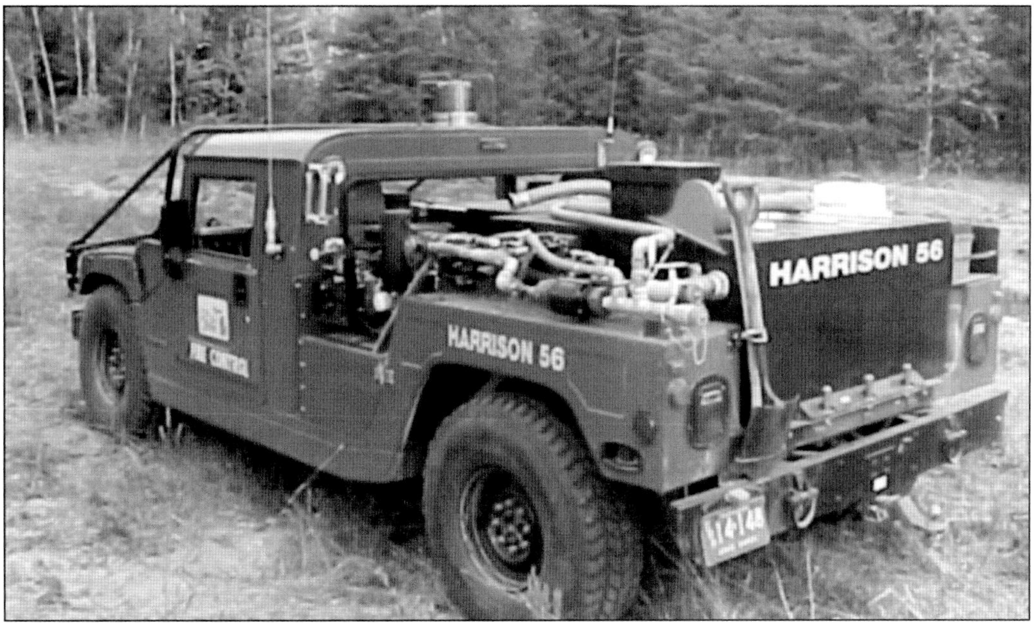

In the 1980s, the AM General Corporation designed the high-mobility, multipurpose, wheeled vehicle (HMMWV) for the military and, later, the similar commercial Hummer vehicle. In 1989, AM General provided REC with a commercial Hummer for field evaluation prior to the unit being available for purchase by the public in 1992. Both the military and commercial vehicles were tested with results and designs published in Project No. 56 A and B. (Michigan Roscommon Equipment Center.)

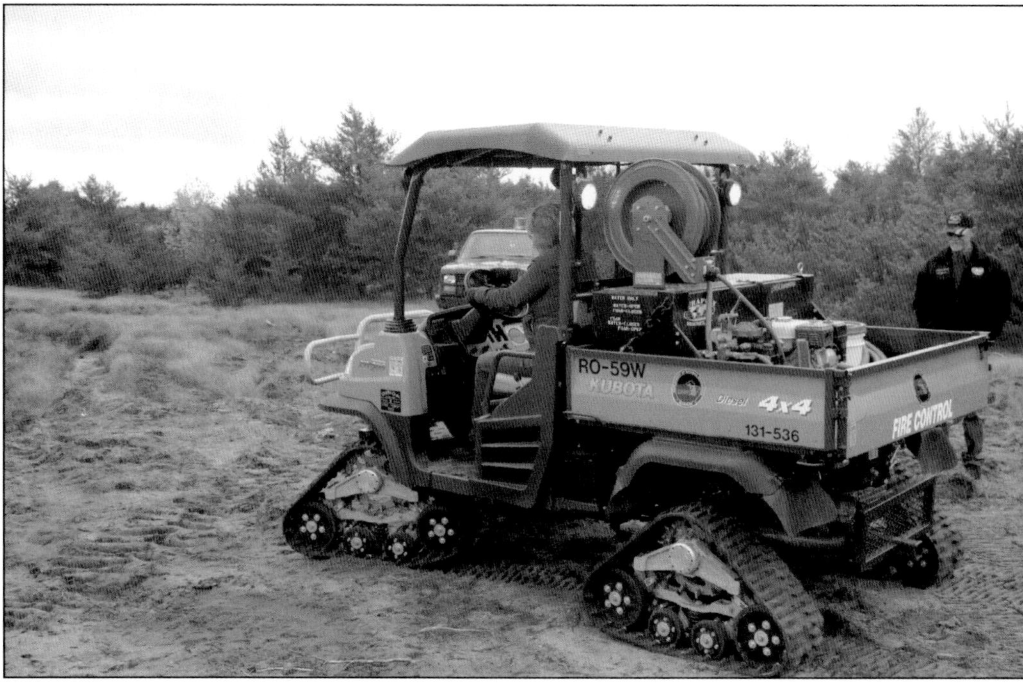

REC Project No. 69 (2009) was developed to identify utility terrain vehicle (UTV) specifications for use by the wildland fire community. REC found that these units make fire suppression operations more efficient, but doing it safely is the primary objective of the firefighter. The report was intended to assist agencies in determining which UTV could best fit their needs based on the situations of how and where the unit will be used. (Michigan Roscommon Equipment Center.)

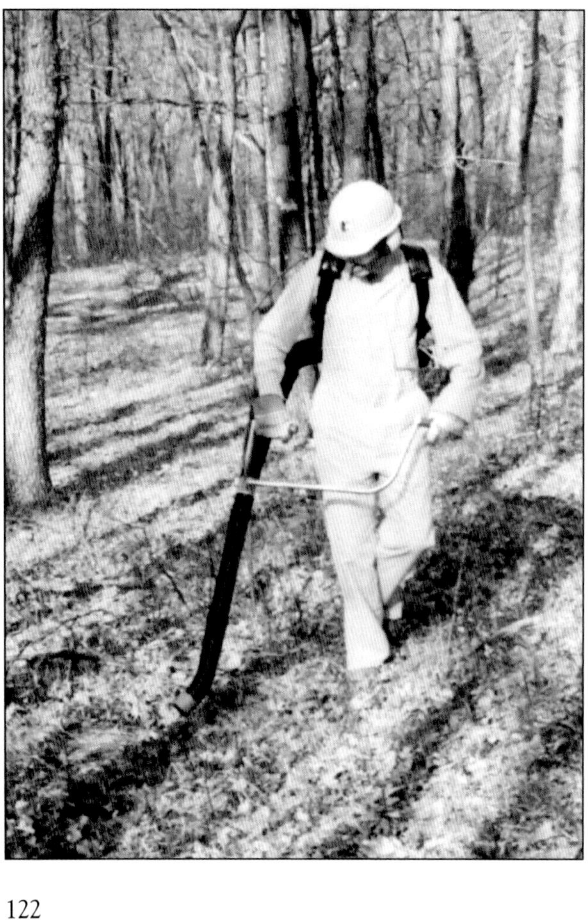

In 1992, the Missouri Department of Conservation initiated a project with REC to study and evaluate power blowers used in constructing fire lines. In the more rugged terrain of the Appalachians and Ozarks within the 20 Northeast states, fire plows are not suitable for use, but blowers are effective in constructing fire lines in hardwood leaf litter. Nine states and several volunteer fire departments participated. (Michigan Roscommon Equipment Center.)

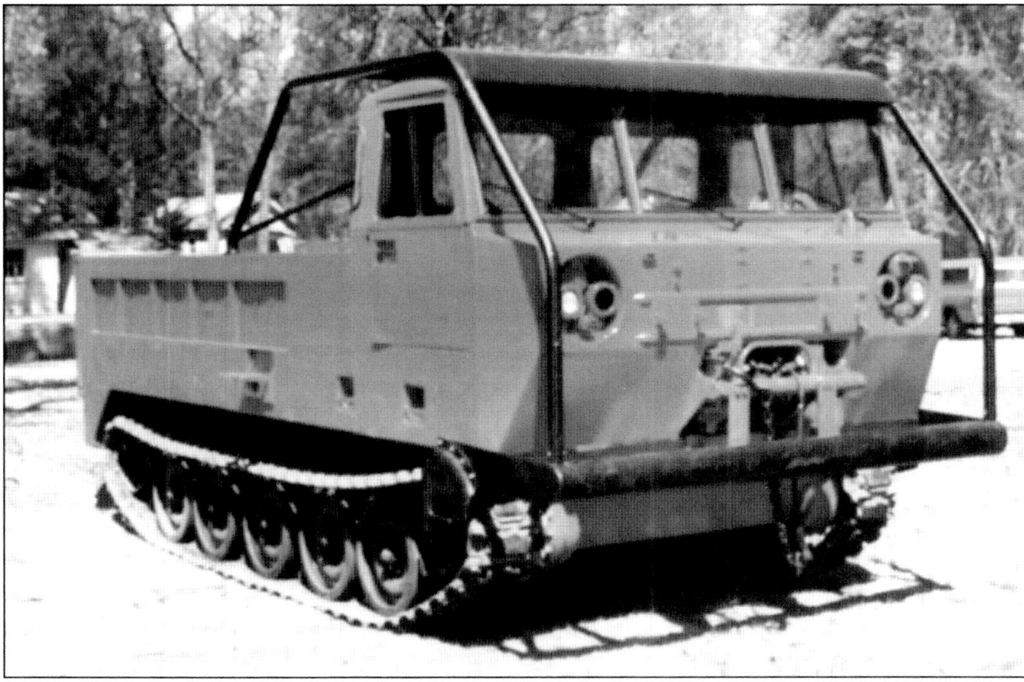

In 1998, REC worked with the states of Maine, North Carolina, and Arkansas along with the Department of Defense at Fort Polk, Louisiana, to evaluate the military M548 and M105 full-tracked vehicle for use in wildland fire suppression. It was determined that the vehicle has limited use in selected locations, such as applications in marshes or bogs where low ground pressure is required or rugged terrain. Limitations included a top road speed of 33 miles per hour, as well as the requirement of a lowboy trailer needed for its transportation. (Michigan Roscommon Equipment Center.)

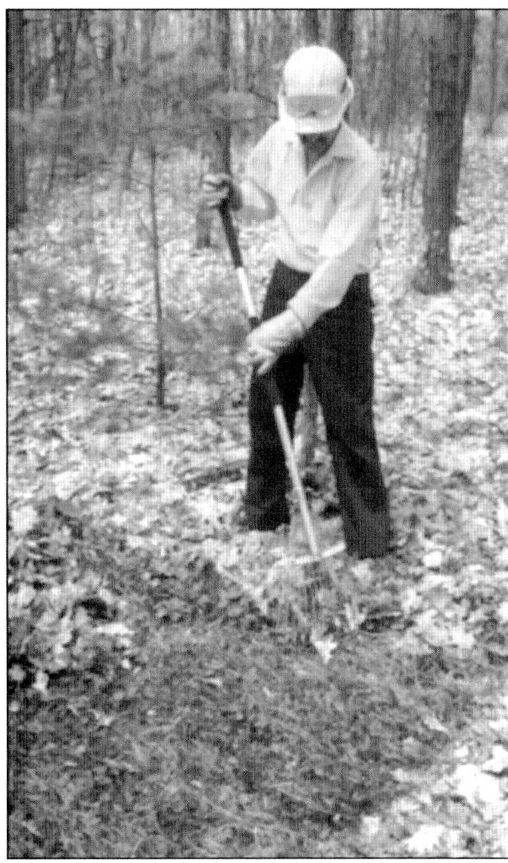

The Collapsible Fire Rake project was initiated by Indiana and New Jersey in 1995 due to no acceptable rakes being available for purchase from vendors. Eight states participated in testing prototype rakes during the 1996 spring fire season. In 1997, an acceptable rake was developed, and REC met with General Services Administration to request its inclusion in the GSA *Wildland Fire Equipment* catalog. In 1998, GSA issued a contract for 3,000 rakes, and they were available in their spring catalog for $38.26 each. (Michigan Roscommon Equipment Center.)

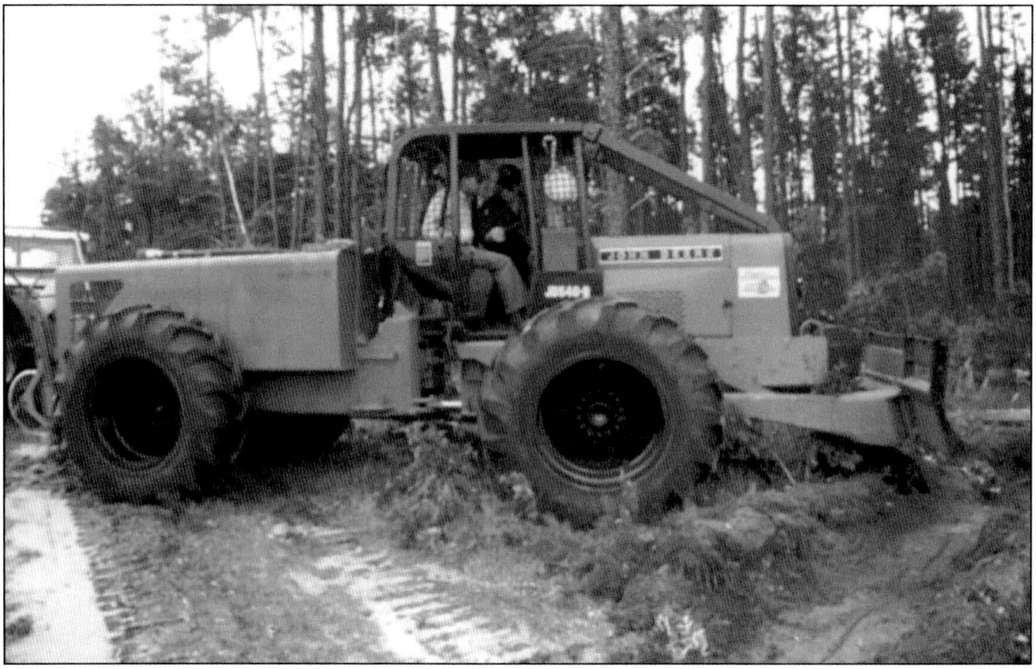

REC completed projects on the use of skidders in 1979, 1988, and 1993. Skidders can use a fixed water tank or a portable tank system. Portable tanks are winched into the arch area of the machine or lifted in place by a grapple. This photograph shows one of Michigan's large skidder with fixed tank. Some skidders have also had fire line plows installed for added capability. (Michigan Roscommon Equipment Center.)

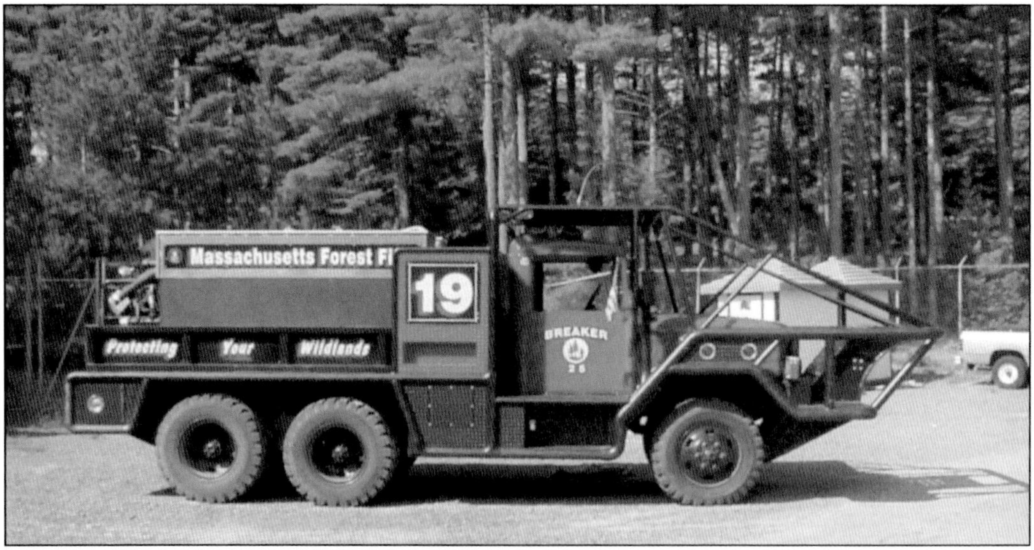

Southeast Massachusetts and Cape Cod have been using armored off-road engines since the 1930s. They are better known as brush breakers. These units have distinctive armoring that allows them to push over pines and small oaks in pine barren vegetation, which allows a mobile, direct attack on fast-spreading fires. Breaker No. 19 is a 1998 V-Tech built on a 1975 AM General Motors military chassis acquired through the Federal Excess Personal Property program and is stationed at Myles Standish State Forest. (NFFPC.)

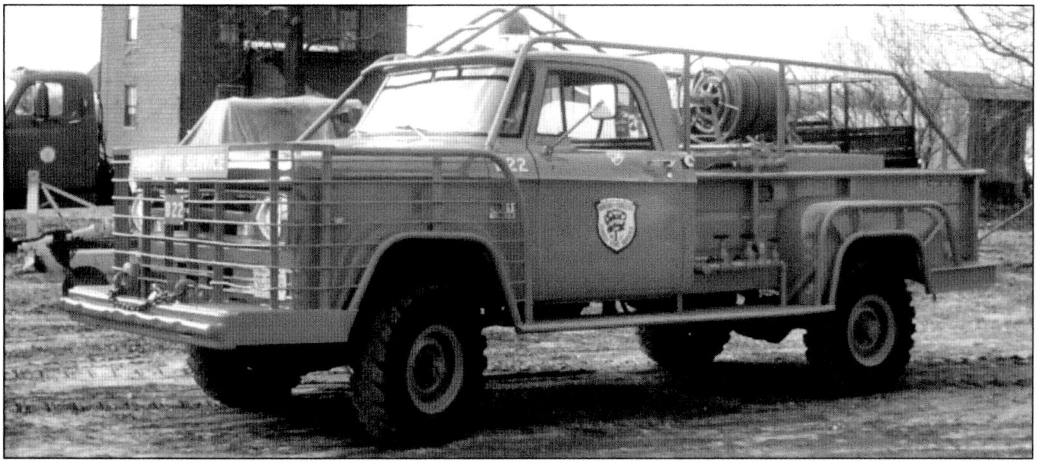

In 1981, REC Project No. 8 was completed, and it included a full set of schematic drawings of brush armoring of a New Jersey Dodge Model W-300 off-road engine. At this time, almost every state in the Northeast was utilizing four-by-four or six-by-six trucks for forest fire control work. Some engines were well guarded against damage from being used off-road while others had no protection at all. This project was completed in order to share the armoring plans among the states so that they could develop brush armoring for the engines that they utilized. (New Jersey Forest Fire Service.)

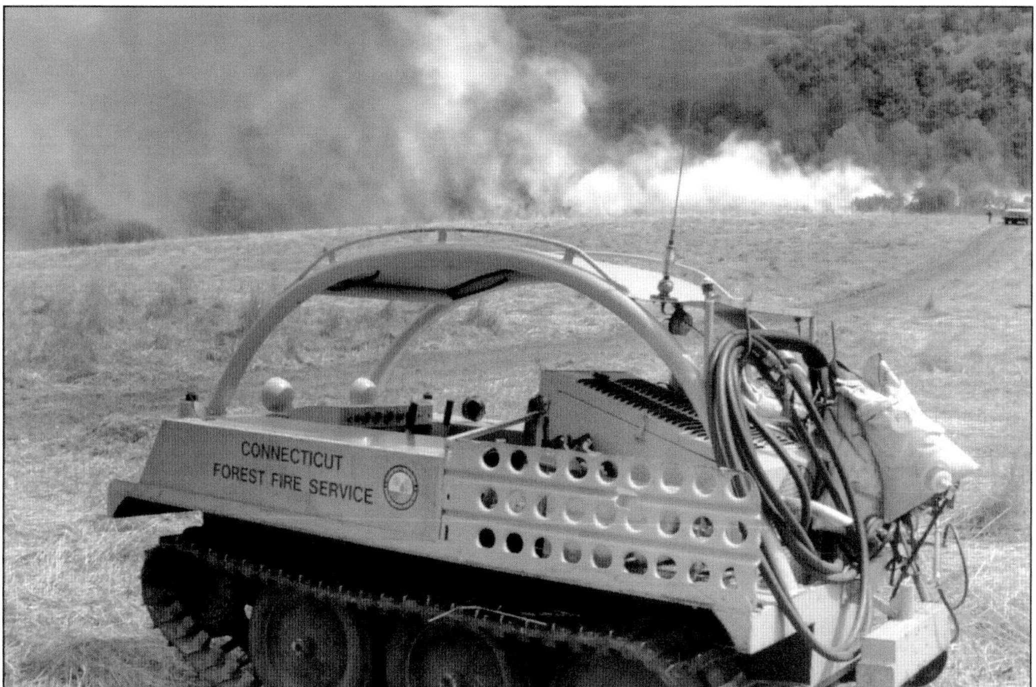

REC Project No. 13 evaluated the use of the Bombardier J-5 in wildland firefighting. At this point, Connecticut, Kentucky, Maine, Minnesota, New Jersey, and Wisconsin were utilizing these units and were included in the evaluation process. Minnesota had 21 machines in service while the Kentucky and Wisconsin units were equipped with fire line plows. The J-5's ability to negotiate bogs and swamps was identified as one of its best features. This photograph shows a CT J-5 standing by at a prescribed burning project. (Connecticut Division of Forestry.)

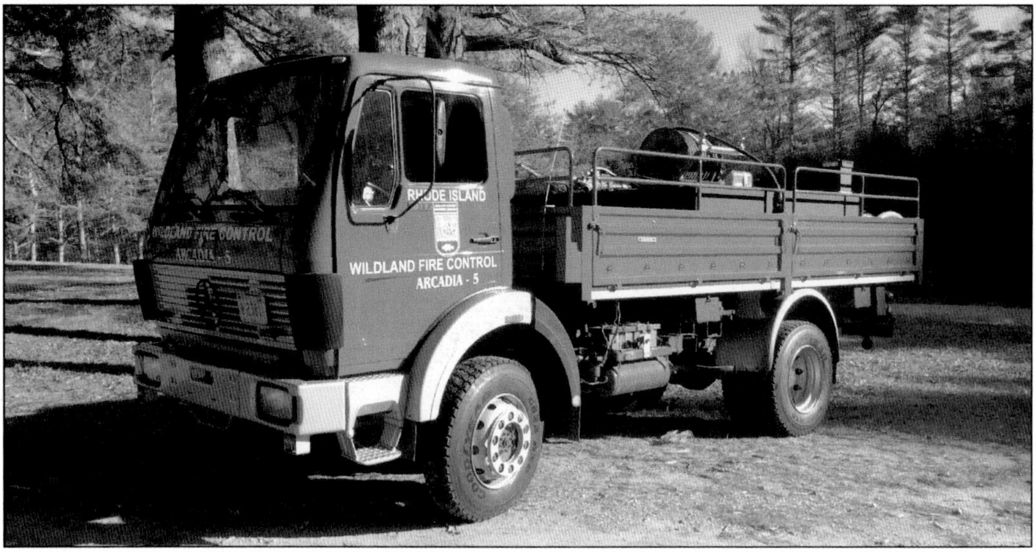

The Mercedes-Benz 1017 truck was originally designed for the West German Army and was constructed from 1977 through 1988. It is essentially a commercial vehicle with minimum modifications to meet military requirements. One thousand seventeen is short for the truck's maximum gross weight of 10 tons, and its engine has a horsepower of 170. The Rhode Island Division of Forest Environment acquired this truck through the Federal Excess Personal Property program in 2001 and has converted it to a wildland engine with an 800-gallon water tank. (Rhode Island Division of Forest Environment.)

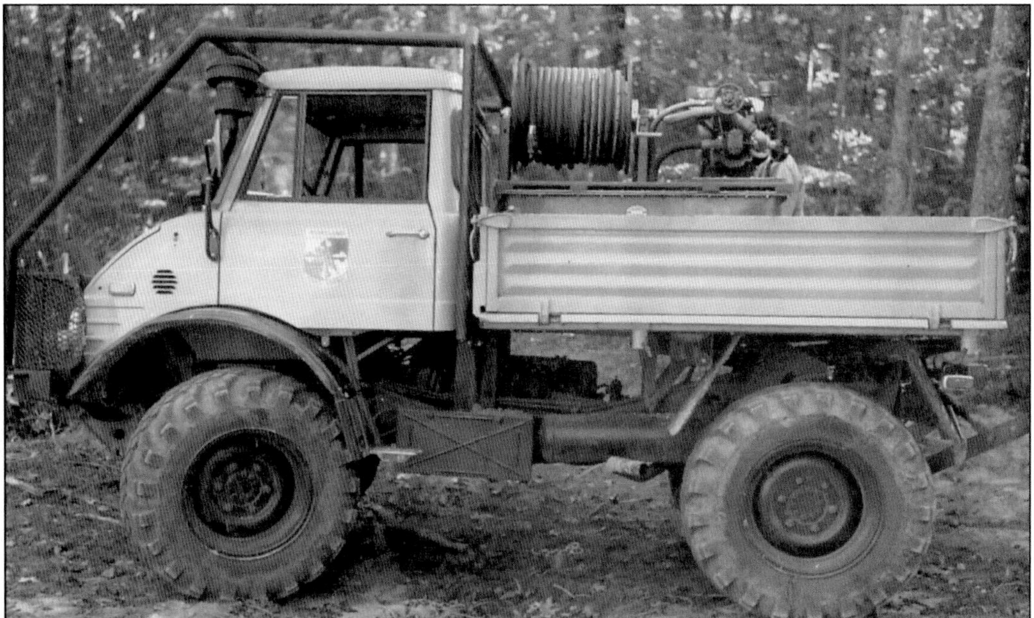

The Maryland Forest Service utilized this unimog on its Eastern Shore during the 1970s and 1980s. In 1979, the Roscommon Equipment Center completed an evaluation of the unimog to determine its capabilities in controlling wildland fires with a water package, as well as with a fire line plow. At the time, Mercedes-Benz advertised that a total of 84 attachments had been developed for use with the unimog. (Maryland Forest Service.)

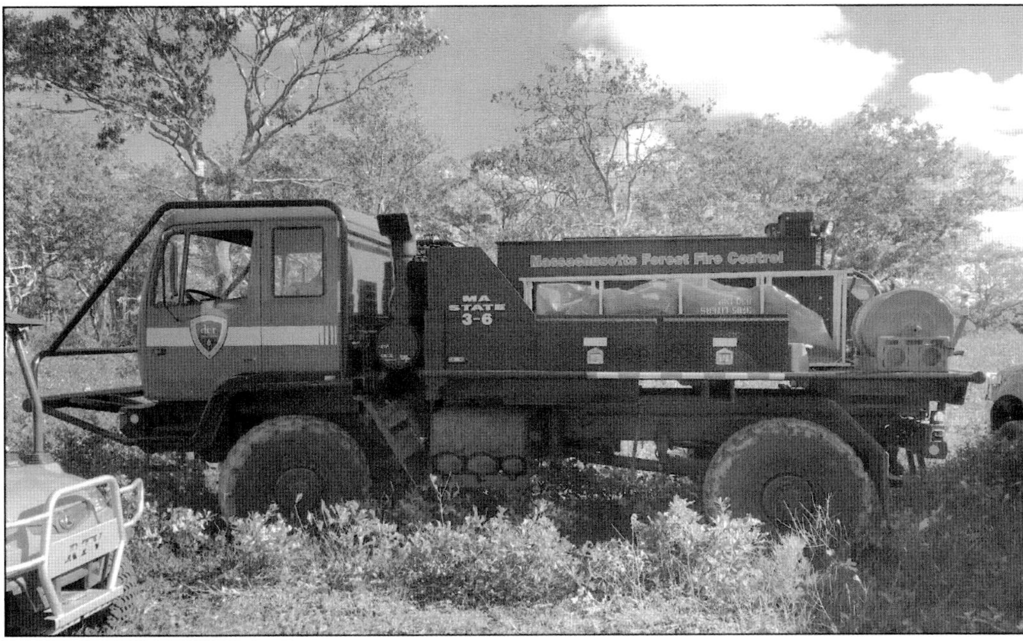

Massachusetts Department of Conservation and Recreation's Bureau of Forest Fire Control's Engine 3-6 was built from a former military Stewart Stevens M1078. This 2016 photograph captures its first assignment supporting a prescribed fire project on Martha's Vineyard. Acquired through the US Forest Service and the Department of Defense's Firefighter Property program, it is configured with a 600-gallon slip-on unit. This type of unit is slowly becoming the new generation 2.5-ton off-road engine. (Massachusetts Bureau of Forest Fire Control.)

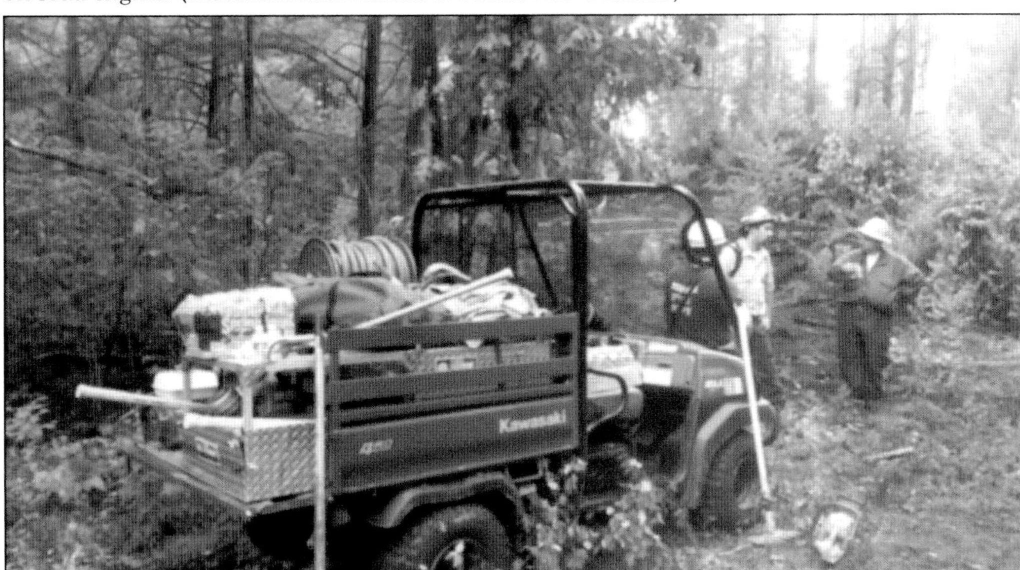

This UTV, owned and operated by a Massachusetts volunteer fire department, is shown on a remote four-acre wildfire where access was very limited. The vehicle is outfitted with a 65-gallon slip-on unit funded through the US Forest Service and Massachusetts Department of Conservation and Recreation's Bureau of Forest Fire Control's Volunteer Fire Assistance program. (Massachusetts Bureau of Forest Fire Control.)